Smartphones as Mobile Minilabs in Physics

Jochen Kuhn • Patrik Vogt

Editors

Smartphones as Mobile Minilabs in Physics

Edited Volume Featuring more than 70 Examples from 10 Years *The Physics Teacher*-column iPhysicsLabs

 Springer

Editors
Jochen Kuhn
Ludwig-Maximilians-Universität
München (LMU Munich)
Faculty of Physics, Chair of Physics
Education
Munich, Germany

Patrik Vogt
Institute of Teacher Training (ILF) Mainz
Mainz, Germany

ISBN 978-3-030-94043-0   ISBN 978-3-030-94044-7   (eBook)
https://doi.org/10.1007/978-3-030-94044-7

This Springer imprint is published by the registered company Springer Nature Switzerland AG
The registered company address is: Gewerbestrasse 11, 6330 Cham, Switzerland

Foreword

Swiss Pocket knife, indeed! This metaphor might need some explanation for some, but it does seem right on target to me, so please indulge me while I reminisce a bit. In my younger days, I kept an actual Swiss Pocket knife with me all the time, and it seemed a critical part of the prepared physics teacher's equipment—I found myself using my it most every day to cut string or duct tape, or to open boxes or dig holes or fashion cardboard lens holders or any number of other essential physics teacher tasks. But then 9/11 happened and if you traveled much at all, you couldn't risk carrying your Swiss Pocket knife on your person and having it confiscated and yourself detained, so it went into back drawer somewhere and saw much less use than before. Now I no longer carry a knife in my pocket, but the essential piece of physics teacher equipment that has replaced it in many regards, is the cell phone, at least in my case. It is more versatile than the knife in many regards, but in my ideal world I'd have them both at my fingertips when teaching.

When Jochen Kuhn and Patrik Vogt first proposed a column for *The Physics Teacher* on the topic of using cell phones in the classroom, I can imagine there were many skeptics—after all, some teachers were still not allowing calculators in the classroom, if I remember correctly! Their first column was a beautiful piece[1] about using a cell phone to detect diffraction from an infrared remote-control device, inspiringly simple and satisfyingly packaged, but could there be enough other good ideas to generate a quality article every month? It seems even the *TPT* editor at the time, Karl Mamola, was unsure whether this idea would be sustainable as a regular *TPT* feature.

In fact, the "Column Editor's Note" that served to introduce this new *TPT* feature to readers included this caveat, "We will publish 'iPhysicsLabs' on a trial basis for a number of months; depending on reader response it may evolve into a regular *TPT* column."

However, even back in February 2012, when this first column appeared in *TPT*, Kuhn and Vogt seemed to know that this bit of technology was to be a key part of the physics teacher's tool kit for many years to come. In describing the kinds of submissions they hoped to receive, this wildly optimistic pair of physicists had

---

[1] Phys. Teach. **50**, 118 (2012); https://doi.org/10.1119/1.3677292

clear ideas about what would work best, writing: "The column will feature short papers (generally less than 1000 words) describing experiments that make use of the sophisticated features of mobile media devices produced by various manufacturers ... The contributions should include some theoretical background, a description of the experimental setup and procedure, and a discussion of typical results." Kuhn and Vogt have stayed true to this original vision, describing many dozens of clever physics experiments described within the column, and providing teachers everywhere with a host of options for tomorrow's classroom. Some ideas came from Kuhn and Vogt directly while many others were contributed by *TPT* readers—apparently readers were so responsive that the column is still going strong after 10 years! ... And such variety! Topics range from acoustics to mechanics to optics to electromagnetism to radioactivity to thermodynamics to astronomy to almost anything that an introductory physics teacher could imagine wanting to present in their classroom.

Their column has become one of the most long-running columns in *TPT*, being downloaded many times by teachers all over the world within a short period after appearing. In late 2013, when I inherited *The Physics Teacher* editorial role from Karl Mamola, the iPhysics column was firmly established as a regular feature with Kuhn and Vogt as Column Editors. I am so glad that they persisted these past 10 years to produce such a powerful collection of physics experiments that can be done with a pocket device. To have them all in one place, complete with an introductory piece to flesh out some of the logic behind why mobile phone media can be a key part to effective physics pedagogy, is a real benefit to the physics teaching community. Thank you, Jochen and Patrik, for sharing your knowledge and experiences in this way, and congratulations on the tenth anniversary of a truly valuable enterprise!

American Association of Physics Teachers   Gary White
College Park
MD, USA

Contents

Part I

Introduction

Smartphones and Tablet PCs: Excellent Digital Swiss Pocket Knives for Physics Education

1

Jochen Kuhn and Patrik Vogt

Smartphones and tablet PCs are increasingly part of our everyday life—for both the younger and the older generation. Tablet PCs are also increasingly being used in schools, although such devices have primarily been used so far as a substitute for notebooks (e.g., as cognitive tools). Smartphones, however, pose several problems in everyday school life, since for instance they can be distracting or cause disturbances. However, considering their technical possibilities, and learners' deep familiarity with the devices, targeted use of these technologies has long been identified as a possible means to enrich lessons [1].

In addition to the frequently described uses of these technologies, such as for research, as cognitive tools, or for communicating, they can also be used as experimental tools, especially in science classes. Inspired by the early article of Raymond F. Wisman and Kyle Forinash [2], we took our own first steps in this direction in 2009/2010 [3, 4] and recognized very quickly the tremendous opportunities in this idea. Following this, we wanted to extend this "lab in the pocket" idea by integrating it into well-established learning theories and systematically studying these opportunities across a broad range of physics topics. Although we did not set out to found a new direction, we wanted to establish this initiative through collaboration by inviting as many colleagues as possible to join with us. This led us to the idea for starting the "iPhysicsLabs" column in *The Physics Teacher* [5]. And now, after 10 successful years and some breakthrough ideas delivered by different colleagues, it has been a pleasure for us to have brought together so many colleagues from all over the world to work on this fruitful topic. What is more, it seems that we are just at the

J. Kuhn (✉)
Ludwig-Maximilians-Universität München (LMU Munich), Faculty of Physics, Chair of Physics Education, Munich, Germany
e-mail: jochen.kuhn@lmu.de

P. Vogt
Institute of Teacher Training (ILF) Mainz, Mainz, Germany
e-mail: vogt@ilf.bildung-rp.de

J. Kuhn, P. Vogt (eds.), *Smartphones as Mobile Minilabs in Physics*,
https://doi.org/10.1007/978-3-030-94044-7_1

beginning of our journey, as more and more colleagues engage with our idea by developing their own respective apps, such as the Physics Toolbox from Vieyra Software in 2013, phyphox from RWTH Aachen in 2016 [6], and PocketLab, as well as building further communities, such as SmarterPhysics by Martín Monteiro et al. and the Smarte Physik column [7].

Therefore, in this chapter, we give a summary of the basic ideas of this approach, reasons for its promise, and how it can be further developed.

1.1   Mobile Mini-Labs for Teaching and Learning

Mobile communication technologies can be used for a wide range of experiments, especially in physics, as they are equipped with different internal sensors that record physical data. These include, for example, microphone and camera, accelerometer, sensors for magnetic field strength, illumination, or brightness sensors, a gyroscope, GPS receiver, and sometimes even temperature, pressure, and humidity sensors. The original reason why the sensors were installed was of course not purposes to implement them for experiments in science education. The acceleration sensor is used, for example, to determine the device's tilt and to adjust the screen to its orientation. The magnetic field strength sensor is used as a compass to support navigation using the smartphone or to inform the user about position-specific environmental data (temperature, air pressure, humidity, etc.). However, physical data recorded by the internal sensors can be used beyond their actual function with the aid of apps, so that both qualitative and quantitative experiments are possible across a wide range of subject areas, and particularly for physics lessons. Smartphones and tablet PCs thus represent small, portable measurement laboratories that can replace confusing experimental apparatus. Furthermore, they are well known to learners in their everyday lives, which means that a high level of familiarity with their operation can be expected. Many experiments that can be carried out with mobile communication media were previously only possible with the support of computers and sensors, and some of these were expensive and difficult to operate. In contrast, experiments with the internal sensors of smartphones or tablet PCs can be carried out and evaluated more easily due to the intuitive usability of the apps, so that a stronger focus on physical content is possible.

The papers of this book discuss numerous topics of physics, and its structure corresponds to that of a typical standard work on experimental physics. We start with kinematics and dynamics (Part II), such as with ball velocity, free fall, and the Atwood machine. The reader will find impact processes in Part III, before the topic of rotation (Part IV) is addressed in detail. Topics include the direction of acceleration, angular momentum, and the SpillNot, which enables full glasses to be transported without spilling. In the following parts, deformable bodies (Part V) and pendulum experiments are discussed (Part VI), before we enter the broad field of acoustics (Parts VII–IX). Among other things, numerous variants for the determination of the speed of sound in gases and solids, as well as various everyday phenomena such as the knuckle cracking, opening wine bottles, and the physics of church bells, are presented here. Some experiments on thermodynamics (Part X), electrodynamics (Part XI), and optics (Part XII), as well as astronomy and modern physics (Part XIII), complete the edited volume.

1.2 Good Reasons for Learning with Mobile Mini-Labs

Apart from the fact that smartphones and tablet PCs are technically and practically suitable for experimental use in physics education, there are also reasons for their meaningful use based on learning theories.

First, the use of the devices as experimental tools in science lessons is didactically justified by the everyday relevance of the smartphone or tablet PC. Thus, it can be classified in well-founded theories of cognitive science, namely situated learning (e.g., [8, 9]) and context-based science education (see, e.g., [10]). The assumption here is that, in addition to the authenticity (in the sense of everyday relevance) of a topic, the authenticity of the media used in experiments also has a positive influence on learning in physics education (so-called material situatedness). Specifically, this assumption means that learners' cognitive and motivational learning success with respect to experiments in physics classes is greater when they investigate a physical phenomenon with experimental media that they use in their daily life [11].

In addition, students are assumed to have an enhanced experience of autonomy when using smartphones and tablet PCs [12, 13]. For example, they can use a tablet PC to independently record a self-selected movement process, directly analyze and evaluate their "own" videos (captured with their own device) using a video analysis app on the same tablet PC, and conduct similar, repetitive, or advanced experiments with a mobile medium outside of school.

In contrast to "conventional" experiments, such mobile technology can also be used to provide multiple representations for students already during and directly after the experimentation (automatic visualizing measurement data as diagrams and tables of values, formulas, vectors or images). The integration and presentation of this multimedia content is done considering of the Cognitive Theory of Multimedia Learning (CTML) [14]. Through active information processing, the coherent use and construction of multiple mental representations is to be promoted. The competent use of these multiple representations such as pictures, diagrams, formulas, and vectors—i.e., the ability to interpret external forms of representation, generate them independently, and switch between different representations flexibly and purposefully [15]—are summarized under the term of (conceptual) "representational competence" [16, 17].

The important role of representational competence for scientific thinking and learning is well documented for science in general [18] and for different individual disciplines (biology [19]; chemistry [20]; physics [21, 22], and mathematics [23, 24]). Representational competence as a prerequisite for the use of multiple representations in terms of domain-specific thinking tools is of high importance for other abilities, e.g., conceptual understanding [25, 26], "construction and reconstruction of meaning" [27], reasoning [28–30], problem solving [16, 30], and creativity

[31]. Against this background, it becomes clear why this competence is discussed for STEM subjects in general as a necessary prerequisite for the formation of deeper understanding [16, 32]. Etkina et al. [33] even mentioned it as the first of seven discipline-specific skills that should be trained.

On the other hand, research findings indicate that competent handling of representations is of considerable difficulty for learners [34, 35]. Empirical evidence for this exists from primary [34] to secondary [36] and university levels [37]. Of course, using the representation possibilities especially of smartphones and tablet PCs only makes sense if students have previously practiced converting measurement data into different forms of representation by hand. In addition, apps often offer forms of representation that are unsuitable for the data in question (e.g., bar charts, where line charts or dot plots would be necessary). This should also be addressed in class.

Since the idea of using smartphones and tablet PCs as experimental tools in the classroom is still relatively new, there are still hardly any published findings on the effectiveness of their use in this way. In the field of physics, however, it is possible to draw on the initial results of some studies published to date on this topic.

A first pilot study on the use of smartphone experiments in physics classes dealt with the topic of acoustics (secondary level 1) [11]. During two-weeks of physics lessons, the classes each worked in groups on four different learning stations with experiments convering topics of beat frequency, types of sound, sound velocity, and sound propagation. The content, scope, and difficulty of the experiments, as well as the instructional materials of the learning stations in the two classes, were identical and differed only in the experimental material used. In order to track the motivation and learning performance of the students, the necessary data were collected directly before and after the intervention as well as five weeks after the completion of this teaching sequence using curricular related tests and questionnaires. It was found that the time course of performance and the learners' knowledge differed significantly between the two groups. In the group of students who worked with smartphone experiments, performance and their self-efficacy expectations were more enhanced and stabilized, respectively, than in the group with conventional experiments. Even though motivation as a whole was not influenced differently, the motivational aspect of "self-efficacy expectation" in particular, which is significant for context-oriented learning, could thus be supported, although it is considered difficult to change.

In a second study in university introductory physics courses in mechanics, positive effects were also found when using tablet PCs for mobile video analysis on the physical concept of understanding in mechanics as well as on the students' self-concept [17, 38, 39].

The third study investigated smartphone experiments in classical mechanics at secondary level 2 (grade 11) using smartphone internal acceleration sensors [40]. Similar to the previous two studies, an quasi-experimental pre-posttest treatment-control group design was used to investigate the effects of smartphone used as experimental tools on interest, curiosity, and learning outcomes. Learners in the smartphone groups showed significantly higher interest in physics after the study. In this regard, especially such learners in this groups who were less interested at the beginning of the study benefited the most. In addition, learners in the smartphone groups showed higher curiosity. No differences in learning performance were found. This means that the use of smartphone experiments can promote interest and curiosity without reducing learning performance.

A fourth study analyzed the learning efficacy of using tablet PC-based video analysis of motion in the subject area of mechanics in secondary level 2 physics classes [41]. Again a quasi-experimental field study was conducted in a pre-posttest design with control and intervention groups. The study included two essential topics of mechanics: uniform motion and accelerated motion. The results demonstrated significantly higher learning performance related to understanding physics concepts through the use of tablet PC-based video analysis compared to traditional instruction in both topics, with the larger effect in the more cognitively demanding topic of accelerated motion [42, 43]. These results could be reproduced by a further study from Hochberg et al. [44].

These first studies on the use of smartphones and tablet PCs as experimental tools did not, of course, allow any general transferable findings, for reasons such as their small number of participants and the limitation of topic reference and addressee group. However, they have provided indications of initial trends and of questions that are still valid and currently being investigated in further studies with a larger number of participants, different groups of addressees (schools and universities) and other subject areas.

1.3 Summary

In this introductory chapter, we described why smartphones and tablet PCs are particularly well suited for experiments in physics education. Furthermore, we explained with reference to established learning theories why students can be expected to learn better with them. Initial studies show positive learning and motivation effects.

Of course, this is only possible if teachers develop suitable instructions and implement them appropriately into their lessons. A logical conclusion from this is that such examples must be integrated even more strongly into teacher training. The following chapters describe numerous tested experiments that can also be used very well for this purpose.

Acknowledgements We would like to thank the American Association of Physics Teachers (AAPT) in general, and *The Physics Teacher* (TPT) journal in particular, for their support and for providing the opportunity to publish and advance the iPhysicsLabs column. A special thanks goes to Gary White, Pam Aycock, and Jane Chambers for their continuous feedback and help.

Likewise, a big thanks goes to Springer Publishing, and especially to all the authors, for the opportunity and their support and contribution to this edited volume.

References

1. West, M., Vosloo, S.: UNESCO Policy Guidelines for Mobile Learning. UNESCO Publications, Paris (2013)
2. Wisman, R.F., Forinash, K.: Science in your pocket. In: Proceedings of the 5th International Conference on Hands-on Science Formal and Informal Science Education (HSCI), pp. 180–187 (September 2008)
3. Vogt, P., Kuhn, J., Müller, S.: Experiments using cell phones in physics classroom education: the computer aided g-determination. Phys. Teach. **49**, 383–384 (2011)
4. Vogt, P., Kuhn, J., Gareis, S.: Beschleunigungssensoren von Smartphones – Möglichkeiten und Beispielexperimente zum Einsatz im Physikunterricht. Praxis der Naturwissenschaften – Physik in der Schule. **60**(7), 15–23 (2011)
5. Kuhn, J., Vogt, P.: iPhysicsLabs. Column editors' note. Phys. Teach. **50**, 118 (2012)
6. Stampfer, C., Heinke, H., Staacks, S.: A lab in the pocket. Nat. Rev. Mater. **5**, 169–170 (2020)
7. Kuhn, J., Wilhelm, T., Lück, S.: Smarte Physik: Physik mit Smartphones und Tablet-PCs. Physik in unserer Zeit. **44**(1), 44–45 (2013)
8. Greeno, J.G., Smith, D.R., Moore, J.L.: Transfer of situated learning. In: Dettermann, D.K., Sternberg, R.J. (eds.) Transfer on Trial: Intelligence, Cognition and Instruction, pp. 99–167. Ablex, Norwood, NJ (1993)
9. Gruber, H., Law, L.-C., Mandl, H., Renkl, A.: Situated learning and transfe. In: Reimann, P., Spada, H. (eds.) Learning in Humans and Machines: Towards an Interdisciplinary Learning Science, pp. 168–188. Pergamon, Oxford (1995)
10. Kuhn, J., Müller, A., Müller, W., Vogt, P.: Kontextorientierter Physikunterricht: Konzeptionen, Theorien und Forschung zu Motivation und Lernen. Praxis der Naturwissenschaften – Physik in der Schule. **59**(5), 13–25 (2010)
11. Kuhn, J., Vogt, P.: Smartphone & Co. in physics education: effects of learning with new media experimental tools in acoustics. In: Schnotz, W., Kauertz, A., Ludwig, H., Müller, A., Pretsch, J. (eds.) Multidisciplinary Research on Teaching and Learning, pp. 253–269. Palgrave Macmillan, Basingstoke, UK (2015)
12. Ryan, R.M., Deci, E.L.: Self-determination theory and the facilitation of intrinsic motivation, social development, and well-being. Am. Psychoanal. **55**, 68–78 (2000)
13. Ryan, R.M., Deci, E.L.: Intrinsic and extrinsic motivations: classic definitions and new directions. Contemp. Educ. Psychol. **25**, 54–67 (2000)
14. Mayer, R.E.: Multimedia Learning, 2nd edn. Cambridge University Press, New York (2009)
15. De Cock, M.: Representation use and strategy choice in physics problem solving. Phys. Rev. Phys. Educ. Res. **8**(2), 020117 (2012)
16. Kohl, P., Finkelstein, N.: Students' representational competence and self-assessment when solving physics problems. Phys. Rev. Phys. Educ. Res. **1**, 010104 (2005)
17. Klein, P., Müller, A., Kuhn, J.: KiRC inventory: assessment of representational competence in kinematics. Phys. Rev. Phys. Educ. Res. **13**, 010132 (2017)
18. Tytler, R., Prain, V., Hubber, P., Waldrip, B. (eds.): Constructing Representations to Learn in Science. Sense, Rotterdam (2013)
19. Tsui, C., Treagust, D. (eds.): Multiple Representations in Biological Education. Springer, Dordrecht (2013)
20. Gilbert, J.K., Treagust, D. (eds.): Multiple Representations in Chemical Education. Springer, Dordrecht (2009)
21. Docktor, J.L., Mestre, J.P.: Synthesis of discipline-based education research in physics. Phys. Rev. Phys. Educ. Res. **10**(2), 020119 (2014)
22. Treagust, D., Duit, R., Fischer, H. (eds.): Multiple Representations in Physics Education. Springer, Dordrecht (2017)
23. Lesh, R., Post, T., Behr, M.: Representations and translations among representations. In: Janvier, C. (ed.) Problems of Representation in the Teaching and Learning of Mathematics, pp. 33–40. Lawrence Erlbaum, Hillsdale (1987)

24. Even, R.: Factors involved in linking representations of functions. J. Math. Behav. **17**, 105–121 (1998)
25. Van Heuvelen, A., Zou, X.: Multiple representations of workenergy processes. Am. J. Phys. **69**, 184 (2001)
26. Hubber, P., Tytler, R., Haslam, F.: Teaching and learning about force with a representational focus: pedagogy and teacher change. Res. Sci. Educ. **40**(1), 5–28 (2010)
27. Opfermann, M., Schmeck, A., Fischer, H.: Multiple representations in physics and science education – why should we use them? In: Treagust, D., Duit, R., Fischer, H. (eds.) Multiple Representations in Physics Education. Springer, Dotrecht (2017)
28. Van Heuvelen, A.: Learning to think like a physicist: a review of research-based instructional strategies. Am. J. Phys. **59**, 891–897 (1991)
29. Plötzner, R., Spada, H.: Constructing quantitative problem representations on the basis of qualitative reasoning. Interact. Learn. Environ. **5**, 95–107 (1998)
30. Verschaffel, L., De Corte, E., de Jong, T., Elen, J.: Use of External Representations in Reasoning and Problem Solving. Routledge, New York (2010)
31. Schnotz, W.: Reanalyzing the expertise reversal effect. Instr. Sci. **38**(3), 315–323 (2010)
32. DiSessa, A.A.: Metarepresentation: native competence and targets for instruction. Cogn. Instr. **22**, 293–331 (3/2004)
33. Etkina, E., Van Heuvelen, A., White-Brahmia, S., Brookes, D.T., Gentile, M., Murthy, S., Rosengrant, D., Warren, A.: Scientific abilities and their assessment. Phys. Rev. Phys. Educ. Res. **2**(2), 020103-1–020103-15 (2006)
34. Ainsworth, S.E., Bibby, P.A., Wood, D.J.: Examining the effects of different multiple representational systems in learning primary mathematics. J. Learn. Sci. **11**, 25–61 (2002)
35. Schoenfeld, A., Smith, J.P., Arcavi, A.: Learning: the microgenetic analysis of one student's evolving understanding of a complex subject matter domain. In: Glaser, R. (ed.) Advances in instructional psychology. LEA, Hillsdale (1993)
36. Scheid, J., Müller, A., Hettmansperger, R., Kuhn, J.: Erhebung von repräsentationaler Kohärenzfähigkeit von Schülerinnen und Schülern im Themenbereich Strahlenoptik. Zeitschrift für Didaktik der Naturwissenschaften. **23**, 181–203 (2017)
37. Nieminen, P., Savinainen, A., Viiri, J.: Force concept inventory-based multiple-choice test for investigating students' representational consistency. Phys. Rev. Phys. Educ. Res. **6**(2), 020109 (2010)
38. Klein, P., Kuhn, J., Müller, A., Gröber, S.: Video analysis exercises in regular introductory mechanics physics courses: Effects of conventional methods and possibilities of mobile devices. In: Schnotz, W., Kauertz, A., Ludwig, H., Müller, A., Pretsch, J. (eds.) Multidisciplinary Research on Teaching and Learning, pp. 270–288. Palgrave Macmillan, Basingstoke, UK (2015)
39. Klein, P., Kuhn, J., Müller, A.: Förderung von Repräsentationskompetenz und Experimentbezug in den vorlesungsbegleitenden Übungen zur Experimentalphysik – Empirische Untersuchung eines videobasierten Aufgabenformates. Zeitschrift für Didaktik der Naturwissenschaften. **24**(1), 17–34 (2018)
40. Hochberg, K., Kuhn, J., Müller, A.: Using Smartphones as experimental tools – effects on interest, curiosity and learning in physics education. J. Sci. Educ. Technol. **27**(5), 385–403 (2018)
41. Becker, S., Klein, P., Gößling, A., Kuhn, J.: Förderung von Konzeptverständnis und Repräsentationskompetenz durch Tablet-PC-gestützte Videoanalyse: Empirische Untersuchung der Lernwirksamkeit eines digitalen Lernwerkzeugs im Mechanikunterricht der Sekundarstufe 2. Zeitschrift für Didaktik der Naturwissenschaften. **25**(1), 1–24 (2019)
42. Becker, S., Gößling, A., Klein, P., Kuhn, J.: Using mobile devices to enhance inquiry-based learning processes. Learn. Instr. **69**, 101350 (2020)
43. Becker, S., Gößling, A., Klein, P., Kuhn, J.: Investigating dynamic visualizations of multiple representations using mobile video analysis in physics lessons: effects on emotion, cognitive

load and conceptual understanding. Zeitschrift für Didaktik der Naturwissenschaften. **26**(1), 123–142 (2020)

44. Hochberg, K., Becker, S., Louis, M., Klein, P., Kuhn, J.: Using smartphones as experimental tools – a follow-up: cognitive effects by video analysis and reduction of cognitive load by multiple representations. J. Sci. Educ. Technol. **29**(2), 303–317 (2020)

Part II

Kinematics and Dynamics

Determining Ball Velocities with Smartphones

2

Patrik Vogt, Jochen Kuhn, and Denis Neuschwander

The use of a smartphone's microphone for quantitative analysis in the field of "acoustics" will be discussed in coming chapters of this book, including the analysis of different sources of sound (Chap. 48) [1], the determination of the speed of sound in various gases (Chap. 43) [2], and the examination of acoustic beat frequency phenomena (Chap. 49) [3]. Acoustic data logging can also be very useful in teaching mechanics, for example, to determine g on the basis of bouncing balls (Chap. 12) [4] or on the basis of the Doppler shift [5] (for an overview, see Ref. 6). This chapter adds further to the applied acoustics repertoire and presents an experiment on the determination of the mean velocity of driven or kicked balls.

2.1 Theoretical Background and Execution of the Experiment

Determining the mean velocity of driven and kicked balls by means of acoustics is not new. It has already been discussed by measuring the speed of a soccer ball using a sound card and an external microphone [7]. The basic principle is easy to understand: a sound signal results from kicking the ball. Then this signal is registered by the microphone with a slight delay. The same effect is valid for the ball's impact on the wall. If the microphone's distance from the kicker is identical to its distance

P. Vogt (✉)
Institute of Teacher Training (ILF) Mainz, Mainz, Germany
e-mail: vogt@ilf.bildung-rp.de

J. Kuhn
Ludwig-Maximilians-Universität München (LMU Munich), Faculty of Physics, Chair of Physics Education, Munich, Germany
e-mail: jochen.kuhn@lmu.de

D. Neuschwander
Gymnasium Altona, Hamburg, Germany
e-mail: denis.neuschwander@gym-altona.de

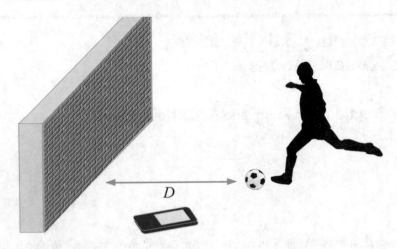

Fig. 2.1 Experimental setup

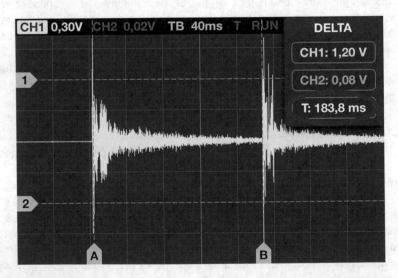

Fig. 2.2 Screen shot of the Oscilloscope App. The first peak shows the impact of the ball on the table-tennis paddle, the second one shows the impact on the wall. The time difference can be found in the app (see red circle)

from the wall, the ball's flight time is the same as the period of time that passes between both registered sound signals (Fig. 2.1). Dividing the person's distance D from the wall by the measured time will reveal the mean velocity of the ball in a very precise way. A smartphone using an oscilloscope app [8] has many advantages over computers, e.g., a higher mobility and a constant availability. The screen shot of the Oscilloscope App in Fig. 2.2 shows the measurement of a driven table-tennis ball with a distance of 2 m from the wall. The mean velocity of 10.9 ms^{-1} results from the actual flight time of 183.8 ms. The following paragraph contains the outcome of several series of measurements. Five different sports have been chosen (table tennis, beach paddle ball, soccer, badminton, and volleyball) to investigate the influence of gender and age on the mean velocity of a driven ball.[1] Calculating the mean value from 10 measurements for each person's performance was necessary in order to gain the highest possible accuracy.

2.2 Experiment Analysis

The results for the different sports are shown in Table 2.1. Typical distances during regular play for each sports activity were used. The maximum mean velocities were achieved in badminton (averaging 81.8 ± 4 km/h). Since the amount of the random sample was relatively small and the subjects were amateurs, the results cannot be considered valid for competitive sports. We saw notable differences between the speeds generated by males and females.

Three soccer teams were separately analyzed in order to examine the influence of age on the mean velocity: a boys' team with group members aged 11–13, a second boys' team with group members aged 15–17, and a men's team with an average age of 24.2 years. As expected, the ball's velocity increases with age in the studied groups (see Table 2.2).

Research shows that studying acoustic phenomena with mobile devices integrated in a more sophisticated instructional setting could also increase learning [9].

Table 2.1 Results for the different sports (N is the sample size, SD is the standard deviation)

Velocity in km/h			
		Male	Female
Sport (distance)	N	11	9
Table tennis	Mean	65.5	44.5
(2 m)	(SD)	(12.5)	(9.5)
Beach paddle ball	Mean	74.7	50.2
(3 m)	(SD)	(13.5)	(10.0)
Soccer	Mean	67.7	42.1
(5 m)	(SD)	(8.9)	(10.4)
Badminton	Mean	81.8	55.7
(3 m)	(SD)	(12.8)	(17.8)
Volleyball	Mean	53.4	35.6
(5 m)	(SD)	(10.7)	(5.9)

---

[1] Due to numerous students playing sports in their free time, they can easily conduct measurements for physics or science classes.

Table 2.2 Velocities in soccer depending on age (*N* sample, *SD* standard deviation)

		Boys' team (11–13 years)	Boys' team (15–17 years)	Men's team
N		9	10	10
Age (mean)		12.2	16.4	24.2
Velocity in km/h	Mean	52.9	79.1	88.5
	(*SD*)	(6.1)	(4.3)	(5.5)

References

1. Kuhn, J., Vogt, P.: Analyzing acoustic phenomena with a smartphone microphone. Phys. Teach. **51**, 118–119 (Feb. 2013)
2. Parolin, S.O., Pezzi, G.: Smartphone-aided measurements of the speed of sound in different gaseous mixtures. Phys. Teach. **51**, 508–509 (Nov. 2013)
3. Kuhn, J., Vogt, P., Hirth, M.: Analyzing the acoustic beat with mobile devices. Phys. Teach. **52**, 248–249 (April 2014)
4. Schwarz, O., Vogt, P., Kuhn, J.: Acoustic measurements of bouncing balls and the determination of gravitational acceleration. Phys. Teach. **51**, 312–313 (May 2013)
5. Vogt, P., Kuhn, J., Müller, S.: Experiments using cell phones in physics classroom education: the computer-aided *g* determination. Phys. Teach. **49**, 383–384 (Sept. 2011)
6. Kuhn, J., Vogt, P.: Smartphones as experimental tools: different methods to determine the gravitational acceleration in classroom physics by using everyday devices. Eur. J. Phys. Educ. **4**(1), 16–27 (2013)
7. Aguiar, C.E., Pereira, M.M.: Using the sound card as a timer. Phys. Teach. **49**, 33–35 (Jan. 2011)
8. https://ogy.de/oscilloscope
9. Kuhn, J., Vogt, P.: Smartphone & Co. in physics education: effects of learning with new media experimental tools in acoustics. In: Schnotz, W., Kauertz, A., Ludwig, H., Müller, A., Pretsch, J. (eds.) Multidisciplinary Research on Teaching and Learning, pp. 253–269. Palgrave Macmillan, Basingstoke, UK (2015)

An Experiment of Relative Velocity in a Train Using a Smartphone

3

Aan Priyanto, Yusmantoro, and Mahardika Prasetya Aji

When we travel in a train moving at a certain velocity, we observe the stationary objects outside are moving backwards. These stationary objects seem to move due to a relative velocity [1, 2]. Consider that the stationary object outside the train is a man standing on the stationary floor watching a woman moving on a train. The woman on a train will see the man moving backward with a similar velocity as her. In kinematics, this magnitude of relative velocity will always be the same whether the man is far away from or near the train, as long as he stands on the stationary floor [1, 2].

But in reality, an observer inside the train will see that each outside object has different velocities according to its distance to the train. The closer the object to the train, the larger the relative velocity of the object as perceived by the observer's eyes. When the objects are far away from the train, they seem to be moving slower than the nearby object.

Several papers have involved smartphones to support physics experiments [3–7]. This chapter aims to provide a simple method using a smartphone app to analyze and theoretically prove that any stationary object outside a moving train at any distance has a similar relative velocity to the train.

3.1   Methods

There were several components required to perform this experiment:

A. Priyanto (✉)
Department of Physics, Bandung Institute of Technology, Bandung, Indonesia
e-mail: aanpriyanto@students.itb.ac.id

Yusmantoro · M. P. Aji
Department of Physics, Universitas Negeri Semarang, Semarang, Indonesia
e-mail: mahardika@mail.unnes.ac.id

© The Author(s), under exclusive license to Springer Nature Switzerland AG 2022
J. Kuhn, P. Vogt (eds.), *Smartphones as Mobile Minilabs in Physics*,
https://doi.org/10.1007/978-3-030-94044-7_3

17

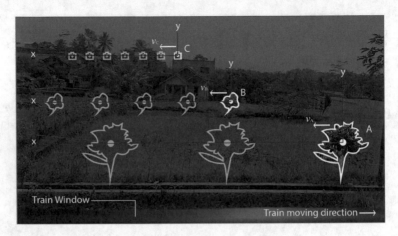

Fig. 3.1 Video shot for objects' motion analysis

1. An Android smartphone capable of recording a video for analysis.
2. VidAnalysis app for Android that can freely be downloaded from the Google Play Store.
3. Microsoft Excel to plot the data and create a graphical analysis.

We required a smartphone to directly record and analyze the motion of several objects outside the train in a certain time as illustrated in Fig. 3.1. For the analysis process, we have made calibration of the distances performed within the VidAnalysis app with a known length placed in the path of the object movement. The resulting time and x-distance data of the objects' movement can then be exported to Excel for graphical process.

3.2   Results and Conclusion

The result of video analysis using a smartphone and graphical analysis using Excel is shown in Fig. 3.2.

The three objects shown in Fig. 3.2 have almost similar velocities. They have about 28.4 m/s of average relative velocity to the train with a mean absolute deviation 0.184. Theoretically, this velocity is also the moving train's velocity. Consider the objects A, B, and C outside the train moving with a velocity of v_T as illustrated in Fig. 3.3.

Figure 3.3 shows the position vector of objects A, B, and C expressed mathematically as in Eqs. (3.1–3.3).

$$r_{TA} = r_T - r_A \tag{3.1}$$

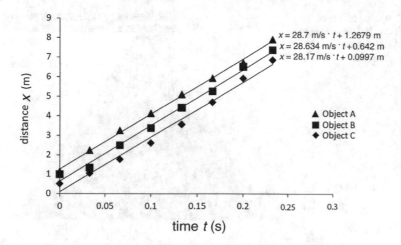

Fig. 3.2 Time (in seconds) and distance (in meters) plot of three stationary objects located at various distances from the moving train

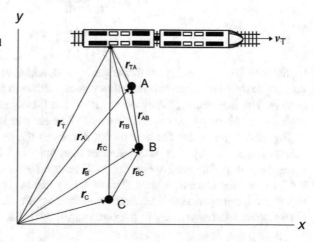

Fig. 3.3 Vector analysis to theoretically find the relative velocities of objects A, B, and C outside a moving train

$$r_{TB} = r_T - r_B \tag{3.2}$$

$$r_{TC} = r_T - r_C \tag{3.3}$$

While the train is moving, position vectors r_A, r_B, and r_C are constant to the origin; thus, we have Eqs. (3.4–3.6):

$$v_{TA} = \frac{dr_T}{dt} + \frac{dr_A}{dt} = \frac{dr_T}{dt} \tag{3.4}$$

Fig. 3.4 Illustration of objects A, B, and C as they pass through the observable area

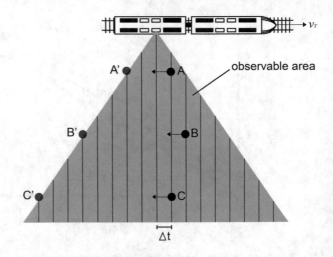

$$v_{\text{TB}} = \frac{\mathrm{d}r_{\text{T}}}{\mathrm{d}t} + \frac{\mathrm{d}r_{\text{B}}}{\mathrm{d}t} = \frac{\mathrm{d}r_{\text{T}}}{\mathrm{d}t} \tag{3.5}$$

$$v_{\text{TC}} = \frac{\mathrm{d}r_{\text{T}}}{\mathrm{d}t} + \frac{\mathrm{d}r_{\text{C}}}{\mathrm{d}t} = \frac{\mathrm{d}r_{\text{T}}}{\mathrm{d}t}. \tag{3.6}$$

Equations (3.4–3.6) confirmed that objects A, B, and C each have a similar relative velocity to the train. There is no relative velocity between the near objects and the far objects. The question is, "Why do the near and far objects seem to have different velocity perceived by the observer's eyes?" The answer is illustrated in Fig. 3.4.

Figure 3.4 shows that the near object A crosses the observable area sooner than objects B and C. Object B passes sooner than object C. Object A requires $3 \cdot \Delta t$ to leave the observable area, while object C as the farthest object needs more time, $9 \cdot \Delta t$, to pass the observable area. The observer inside the train will see the farthest object in a longer period of time than the nearest object. However, all the objects still have similar relative velocity to the observer in the train.

Acknowledgments This work was sponsored by Lembaga Pengelola Dana Pendidikan (LPDP) Republik Indonesia.

References

1. Serway, R.A., Jewett, J.W.: Physics for Scientists and Engineers (with PhysicsNOW and InfoTrac), 6th edn. Brooks Cole, Belmont, CA (2003)
2. Walker, J., Halliday, D., Resnick, R.: Fundamentals of Physics. Wiley, Hoboken, NJ (2011)
3. Tuset-Sanchis, L., Castro-Palacio, J.C., Gómez-Tejedor, J.A., Manjón, F.J., Monsoriu, J.A.: The study of two-dimensional oscillations using a smartphone acceleration sensor: Example of Lissajous curves. Phys. Educ. **50**, 5 (2015)

4. Vogt, P., Kuhn, J., Müller, S.: Experiments using cell phones in physics classroom education: the computer-aided *g* determination. Phys. Teach. **49**, 383 (Sept. 2011)
5. Kuhn, J.: Relevant information about using a mobile phone acceleration sensor in physics experiments. Am. J. Phys. **82**, 2 (Jan. 2014)
6. Azhikannickal, E.: Sports, smartphones, and simulation as an engaging method to teach projectile motion incorporating air resistance. Phys. Teach. **57**, 308 (May 2019)
7. Becker, S., Klein, P., Kuhn, J.: Video analysis on tablet computers to investigate effects of air resistance. Phys. Teach. **54**, 440–441 (Oct. 2016)

LED Gates for Measuring Kinematic Parameters Using the Ambient Light Sensor of a Smartphone

4

Witchayaporn Namchanthra and Chokchai Puttharugsa

Nowadays, electronic devices (especially smartphones) are developed to use as an alternative tool for recording experimental data in physics experiments. This is because of the embedded sensors in a smartphone such as the accelerometer, gyroscope, magnetometer, camera, microphone, and speaker. These sensors were used in physics experiments, such as the simple pendulum, spring pendulum, coupled pendulum, circular motion, Doppler effect, and spectrum of light, as well as video motion analysis (Chaps. 10, 27 and 32) [1–7]. This chapter describes a method for measuring the position-time of a moving object using the ambient light sensor of a smartphone. In the experiment, light-emitting diodes (LEDs), called LED gates, are arranged with equal distance on an inclined plane. A smartphone is attached to a cart and acts as the moving object on the inclined plane. With the appropriate application, the ambient light sensor of the smartphone measures the intensity of the LEDs with respect to time while the smartphone moves along the inclined plane. The results obtained using the LED gates method are in good agreement with that obtained using the video analysis method.

4.1 Theoretical Background and Experimental Setup

The experimental setup consists of an LED strip, a cart with an attached smartphone, and an inclined plane. White LEDs are inserted into a 1.7-m wood strip with equal intervals of 0.1 m. A smartphone (Samsung, Galaxy S4) is attached to the cart using a plastic holder. The cart is made from an aluminum rail (20 cm) with four rubber wheels (see inset in Fig. 4.1). The cart is inserted into the aluminum rail (2.0 m) and is constrained to move on the inclined plane at 5.0°. The smartphone, which has the phyphox app, is used to record the light intensity of the LEDs with respect to time for

W. Namchanthra (✉) · C. Puttharugsa
Department of Physics, Faculty of Science, Srinakharinwirot University, Bangkok, Thailand
e-mail: ped@ioppublishing.com; chokchai@g.swu.ac.th

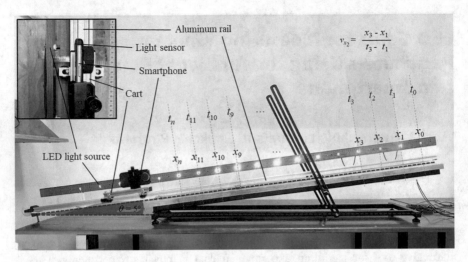

Fig. 4.1 Digital image of the experimental setup (the inset shows a top-view image of the cart with an attached smartphone)

the moving object. The experimental data are emailed to the user for analysis. Moreover, we use the video analysis method by shooting a video of the moving cart using an iPhone 7S (at a frame rate of 30 frames per second) and performing an analysis using the Tracker software to make a comparison with the LED gate method. The position-time data are then plotted and used to calculate the average velocity, which is obtained by using the equation

$$v_n = \frac{\Delta x_n}{\Delta t_n} = \frac{x_{n+1} - x_{n-1}}{t_{n+1} - t_{n-1}}, \tag{4.1}$$

where v_n is the average velocity at position n of LEDs, Δx_n is the position interval n of LEDs, and t_n is the time interval n of LEDs for the cart moving on the inclined plane ($n = 1, 2, 3, \ldots, 17$); x_n is constant and equal to 20 cm.

4.2 Experimental Analysis

Figure 4.2 shows an example of experimental data obtained using the ambient light sensor of a smartphone. It plots the value of illuminance as a function of time for the moving cart. When the cart passes through the LED, the peaks of intensity of the LED appear. The time interval is measured at the middle of the peaks. Thus, we can measure the position-time from the peaks of the cart passing through the LED gates and calculate the average velocity of the cart at position n of the LED using Eq. (4.1).

Based on the experimental data (number of experiments = 3 times), the position-time and average velocity-time plots were drawn as seen in Figs. 4.3(a) and (b) (the black open square), respectively. In Fig. 4.3(b), the cart moved with the increase of

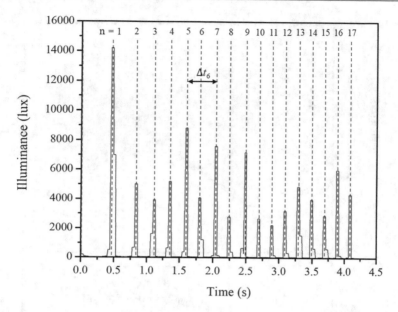

Fig. 4.2 An example of experimental data is recorded through the ambient light sensor of a smartphone using the phyphox application for an object moving on an inclined plane at 5.0°

average velocity up to $t = 1.5$ s, and the average velocity reached a constant at about 0.50 m/s or terminal velocity. Based on the average velocity-time plot, it seems that there is some retarding force acting on the moving object. This occurs because of the friction force of bearing wheels of the cart. In the case of the retarding force, the equation of motion shows that the moving object can be seen in the appendix [8]. According to Eq. (A2) in the appendix, the fitting of the average velocity-time yields the value of $k = 1.345 \pm 0.008$ s^{-1} and $g \cdot \sin(\theta)/k = 0.502 \pm 0.007$ m/s (the value of $g \cdot \sin(\theta)/k$ is the terminal velocity). Moreover, we use the video analysis method to verify the validity of the results obtained using the LED gates method. The position-time obtained from the video analysis using the Tracker software was plotted as shown in Fig. 4.3(a) (the red open dot). The cart positions were tracked and marked with a red dot using a plastic button (Fig. 4.1). It can be seen that the position-time and the average velocity-time obtained from the LED gates and video analysis methods are closely valued. The fitting value of the average velocity obtained using the video analysis method according to Eq. (A2) yields $k = 1.520 \pm 0.008$ s^{-1} and $g \cdot \sin(\theta)/k = 0.492 \pm 0.002$ m/s. The results suggest that we can use LED gates with the ambient light sensor of a smartphone for tracking the position of a moving object on an inclined plane. The LED gates method will provide the position-time information, and students will process this information to obtain the average velocity. We believe that this proposed method will make students understand more the position-time and average velocity-time plots. We have demonstrated an alternative method for measuring the position-time of a moving object on an inclined plane using LED gates with the ambient light sensor

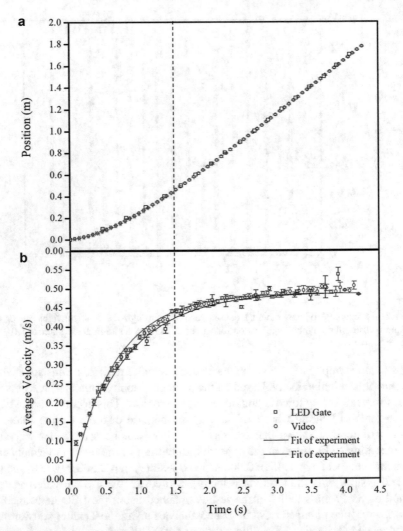

Fig. 4.3 The position-time and the average velocity-time are plotted where the data are obtained from different methods using the LED gates and the video analysis

of a smartphone. The obtained position-time can be calculated for the average velocity of the moving object. The results obtained using the LED gates method correspond to that obtained using the video analysis method.

Acknowledgments This work was supported by the Faculty of Science, Srinakharinwirot University. WN gratefully acknowledges financial support from the Institute for the Promotion of Teaching Science and Technology (IPST) and the National Research Council of Thailand (NRCT).

References

1. Vogt, P., Kuhn, J.: Analyzing collision processes with the smartphone acceleration sensor. Phys. Teach. **52**, 118–119 (Feb. 2014)
2. Kuhn, J., Vogt, P.: Analyzing spring pendulum phenomena with a smartphone acceleration sensor. Phys. Teach. **50**, 504–505 (Nov. 2012)
3. Castro-Palacio, J.C., Velazquez-Abad, L., Gimenez, F., Monsoriu, J.A.: A quantitative analysis of coupled oscillations using mobile accelerometer sensors. Eur. J. Phys. **34**, 737–744 (April 2013)
4. Puttharugsa, C., Khemmani, S., Wicharn, S., Plaipichit, S.: Determination of the coefficient of static friction from circular motion using a smartphone's sensors. Phys. Educ. **54**, 053007 (Sept. 2019)
5. Gomez-Tejedor, J.A., Castro-Palacio, J.C., Monsoriu, J.A.: The acoustic Doppler effect applied to the study of linear motions. Eur. J. Phys. **35**, 025006 (Feb. 2014)
6. Woo, Y., Young-Gu, J.: Fabrication of a high-resolution smartphone spectrometer for education using a 3D printer. Phys. Educ. **54**, 015010 (Jan. 2019)
7. Becker, S., Klein, P., Kuhn, J.: Video analysis on tablet computers to investigate effects of air resistance. Phys. Teach. **54**, 440–441 (Oct. 2016)
8. Readers can see the appendix at TPT Online under the Supplemental tab. https://doi.org/10.1119/10.0004165

Locating a Smartphone's Accelerometer

5

Sidney Mau, Francesco Insulla, Elliot E. Pickens, Zihao Ding, and Scott C. Dudley

While following chapters in this book address aspects of radial acceleration using smartphones (Chaps. 16, 18, 20 and 22) [1–4], this chapter describes a technique to locate the accelerometer in the phone to within a couple millimeters.

5.1 Using a Record Turntable to Determine Accelerometer Location

Centripetal acceleration is given by $a_c = r\omega^2$, with $a_c = \sqrt{a_x^2 + a_y^2}$, where the accelerations are in the rotating smartphone's frame as shown in Fig. 5.1. We measure[1] a_x, a_y, and ω and then calculate the radius as well as an angle θ, given by $\theta = \tan^{-1}(a_y/a_x)$. To accomplish this experimentally, the smartphone is spun at a constant angular speed by attaching it to card stock with a hole positioning it on the turntable's spindle as shown in the inset of Fig. 5.1. If the accelerometer is well calibrated, then spinning the device about a single hole would suffice to determine the sensor's position of the accelerometer inside the phone. This is possible because the magnitude of the centripetal acceleration combined with the angular frequency will yield the radius of a circle on which the sensor should lie. In addition, the ratio of the y- and x-components of the centripetal acceleration yield an angle from the rotation point relative to axes parallel to the sides of the device. For redundancy and confirmation of the method, we checked three rotation points and show the radii and rays for all three in Fig. 5.1. The intersection of the three circles and rays identifies

---

[1] There are many apps, both free and low cost, that allow for recording and exporting of sensor data from a smartphone. For the iPhone we used Sensor Kinetics Pro. For the Android phone we used the Physics Toolbox Apps from Vieyra Software.

S. Mau (✉) · F. Insulla · E. E. Pickens · Z. Ding · S. C. Dudley
TASIS England, Surrey, UK
e-mail: franinsu@stanford.edu; zd75@scarletmail.rutgers.edu; sdudley@tasisengland.org

© The Author(s), under exclusive license to Springer Nature Switzerland AG 2022
J. Kuhn, P. Vogt (eds.), *Smartphones as Mobile Minilabs in Physics*,
https://doi.org/10.1007/978-3-030-94044-7_5

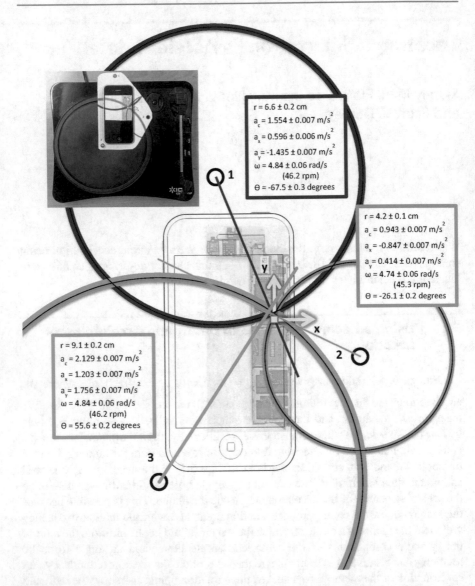

Fig. 5.1 Results for three pivot points about a turntable to determine the location of the accelerometer in an iPhone 4 smartphone. The chip containing the accelerometer is believed to lie at the location of the yellow square (see footnote 2). The inset photo shows the iPhone 4 on the turntable in position 3 with this figure integrated under and superimposed over the phone

the location of the accelerometer's integrated circuit in the phone and provides a visualization of the error in the technique. The method is reminiscent to center of mass determinations by hanging an object and drawing plumb lines, where the lines' intersection is the center of mass.

5.2   Random and Systematic Errors

We found it important to correct the data from the phone for offsets from zero. We achieved this by laying the phone flat for a few seconds and then rotating the phone 180° for another few seconds while taking acceleration measurements. This allowed us to determine the offsets to the x- and y-acceleration measurements. Data for one offset calibration and one pivot point is in Fig. 5.2. All data were taken in a single continuous run, which included offset. Errors reported in the contribution are random errors about the mean [5].

When spinning the phone, the accelerations oscillate due to the turntable being not perfectly level. This is advantageous as it allows for direct measurement of the rotation rate, and does not affect the average value of the accelerations as long as integer numbers of full cycles are averaged. From the measurements we calculate the tilt from horizontal for our setup to be roughly 1°.

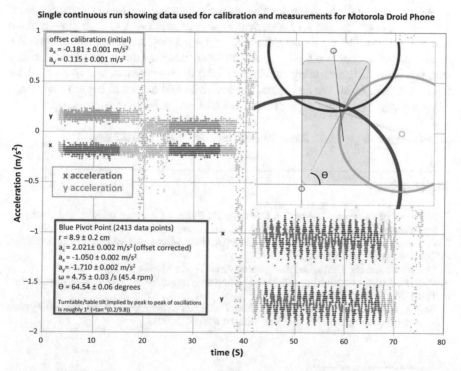

Single continuous run showing data used for calibration and measurements for Motorola Droid Phone

offset calibration (initial)
$a_x = -0.181 \pm 0.001$ m/s²
$a_y = 0.115 \pm 0.001$ m/s²

x acceleration
y acceleration

Blue Pivot Point (2413 data points)
$r = 8.9 \pm 0.2$ cm
$a_c = 2.021 \pm 0.002$ m/s² (offset corrected)
$a_x = -1.050 \pm 0.002$ m/s²
$a_y = -1.710 \pm 0.002$ m/s²
$\omega = 4.75 \pm 0.03$ /s (45.4 rpm)
$\Theta = 64.54 \pm 0.06$ degrees

Turntable/table tilt implied by peak to peak of oscillations
is roughly 1° (=tan⁻¹(0.2/9.8))

Acceleration (m/s²)

time (S)

Fig. 5.2 Data used for radius and angle determination for a single pivot point. Initial offset calibration precedes turntable spinning on the blue pivot point. Points used for calculations are shown in solid colors. Points not used while phone was being positioned are shown as subdued colors. Inset shows results for all three pivot points for this Android phone, the same as that shown in Fig. 5.1 for the iPhone

In Fig. 5.1, the width of the colored circles is the random error in the mean value, which is low since there are much data. The hundreds of points in each measurement allow for good use and discussion of the standard error in the mean.

Even if the random error was reduced to zero, we estimate the systematic errors to still be roughly the width of the colored circles. We see two possible systematic effects. First, the placement and alignment of the phone and transcription of its spatial location, which we did by hand, we estimate to be on the order of a millimeter.[2] Second, we see from the initial and final offset calibrations that there is slight drift even in a single two-minute continuous run. The offsets must be accounted for, and initial frustrations with offsets led us to the single run technique.

5.3   Conclusions

This project provides excellent experience with centripetal acceleration, accelerometers and physics from the sensor's perspective, and error analysis. If 10 years ago one had required a student to acquire and bring to each class a video and still camera, three-axis accelerometers, gyroscopes, magnetometers, an audio recorder, and global positioning unit, one can imagine the costs. Today, many students have such measuring capabilities with them every day and likely will for the rest of their lives. We anticipate that activities and texts will begin to approach physics from the point of view of these sensors, and it is an exciting viewpoint.

Acknowledgments We thank Robert Kennedy for useful discussions and support.

References

1. Monteiro, M., Cabeza, C., Marti, A.C., Vogt, P., Kuhn, J.: Angular velocity and centripetal acceleration relationship. Phys. Teach. **52**, 312–313 (May 2014)
2. Tornaría, F., Monteiro, M., Marti, A.C.: Understanding coffee spills using a smartphone. Phys. Teach. **52**, 502–503 (Nov. 2014)
3. Shakur, A., Sinatra, T.: Angular momentum. Phys. Teach. **51**, 564–565 (Dec. 2013)
4. Vogt, P., Kuhn, J.: Analyzing radial acceleration with a smartphone acceleration sensor. Phys. Teach. **51**, 182–183 (March 2013)
5. Bevington, P.R., Keith Robinson, D.: Data Reduction and Error Analysis for the Physical Sciences, 3rd edn, pp. 60–63. McGraw-Hill, New York (2003)

---

[2]The picture of the circuit board shown in Fig. 5.1 is from https://www.ifixit.com/Teardown/iPhone+4+Teardown/3130 with this image at https://d3nevzfk7ii3be.cloudfront.net/igi/IP1qEpYFQSSqwSbg (fair use per https://www.ifixit.com/Guide/Image/meta/IP1qEpYFQSSqwSbg). To place the accelerometer we noted the dimensions of the A4 chip's package is roughly 7.3 mm each side, and the SIM card slot can be aligned with the opening for same on the side of the phone. But the actual position of this circuit, and thus the yellow square indicating the accelerometer location, still has uncertainties on the order of perhaps 1–3 mm. If we had taken apart our phone we may have improved upon this.

Analyzing Free Fall with a Smartphone Acceleration Sensor

6

Patrik Vogt and Jochen Kuhn

This chapter provides suggestions on how a smartphone can be used to improve mechanics lessons, in particular when used as an accelerometer in the context of laws governing free fall [1]. The app SPARKvue [2] (Fig. 6.1) was used together with an iPhone or an iPod touch, or the Accelogger [3] app if an Android device was used. The values measured by the smartphone were then exported to a spreadsheet application for analysis (e.g., MS Excel).

6.1 Mode of Operation of Acceleration Sensors in Smartphones

It makes sense to fundamentally understand how acceleration sensors work before using them in the classroom. Smartphone acceleration sensors are microsystems that process mechanical and electrical information, so-called micro-electro-mechanical systems (MEMS). In the simplest case, an acceleration sensor consists of a seismic mass that is mounted on spiral springs and can therefore move freely in one direction. If an acceleration a takes effect in this direction, it causes the mass m to move by the distance x. This change in position can be measured with piezoresistive, piezoelectric, or capacitive methods and is a measurement of the current acceleration [4]. In most

P. Vogt (✉)
Institute of Teacher Training (ILF) Mainz, Mainz, Germany
e-mail: vogt@ilf.bildung-rp.de

J. Kuhn
Ludwig-Maximilians-Universität München (LMU Munich), Faculty of Physics, Chair of Physics Education, Munich, Germany
e-mail: jochen.kuhn@lmu.de

Fig. 6.1 Screenshot from the app SPARKvue, showing the setting of the experiment

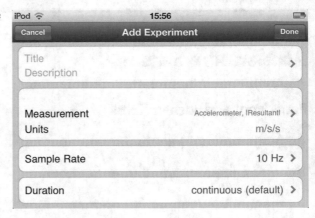

iPod 📶	15:56	🔲
Cancel	**Add Experiment**	Done

Title
Description   >

Measurement	Accelerometer, IResultantI >
Units	m/s/s

Sample Rate   10 Hz >

Duration   continuous (default) >

cases, however, the measurement is made capacitively. Figure 6.2 shows a simplified design of a sensor of this kind: Three silicon sheets, which are placed parallel to each other and connected with spiral springs, make up a series connection of two capacitors. The two outer sheets are fixed; the middle sheet, which forms the seismic mass, is mobile. Acceleration causes the distance between the sheets to shift, leading to changes in capacity. These are measured and converted into an acceleration value. Strictly speaking, they are therefore not acceleration sensors but force sensors.

To measure acceleration three-dimensionally, three sensors have to be included in a smartphone. These sensors have to be positioned orthogonally to each other and determine the acceleration parts a_x, a_y, and a_z of each spatial direction (x-, y, and z-axis) independently (Fig. 6.3).

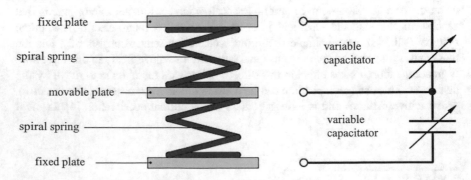

fixed plate

spiral spring

movable plate

spiral spring

fixed plate

variable capacitator

variable capacitator

Fig. 6.2 Design and mode of operation of acceleration sensors [5]

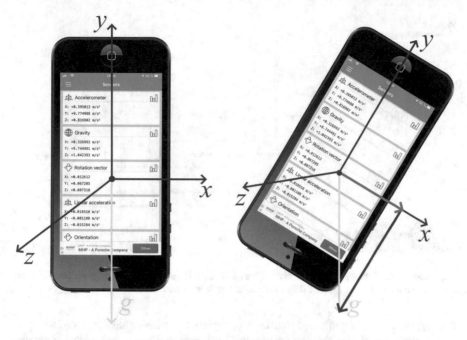

Fig. 6.3 The orientation of the three independent acceleration-sensors of an iPhone or iPod touch; the sensors measure the acceleration in the direction of the three plotted axes

6.2   Study of Free Fall by a Smartphone

A suitable way of examining free fall is to suspend the smartphone from a piece of string, which is burnt through to start the fall [Fig. 6.4(a)]. In order to avoid damaging the device, we place a soft object under the cell telephone (e.g., a cushion) for it to land on. After having started the measurement of acceleration with a measuring frequency of 100 Hz, we burn the string through and the free fall commences. The acceleration value measured can be seen in Fig. 6.4(b).

At first, the smartphone is suspended from the string and the acceleration of gravity of 9.81 ms^{-2} takes effect [left part of Fig. 6.4(b)]. After approx. 0.6 s, the free fall begins and the sensors cannot register any acceleration, because they are being accelerated with 1 g themselves[1]. This state is maintained until the cell phone's fall is stopped by landing on the soft object. As can be seen in Fig. 6.4(b), the sensor

---

[1]This is difficult to understand for pupils because they perceive the exact opposite: At first, the device suspends motionless from a string and then falls, accelerating to the floor. This is why they can only understand the measured acceleration process if they have previously been instructed on the way acceleration sensors function. In addition, the learners' previous experience of being pressed to the floor in a lift accelerating downwards, and the resulting conclusion that one is weightless in a free-falling lift, can also help them understand the process.

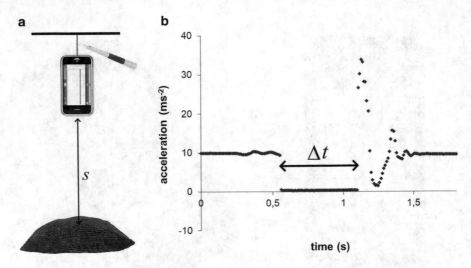

Fig. 6.4 Free fall: (**a**) Experimental setup and acceleration process. (**b**) Presentation of measurements after the export of data from the smartphone into MS Excel

continues to move slightly and returns to complete immobility after a period of 1.5 s. The measurement can then be terminated and exported to a spreadsheet program (e.g., MS Excel) in order to determine the time it takes to fall Δt.

It is obvious that the smartphone has a dual function in this experiment. It serves both as falling body and as electronic gauge, making it possible to determine the free-fall time with a good degree of accuracy. For the measurement example described, the falling time was calculated to be $\Delta t = 0.56$ s for a falling distance of $s = 1.575$ m. If these values are applied to the distance-time equation for uniform acceleration (without initial distance and initial speed and with the influence of the gravitational field for acceleration)

$$s = \frac{1}{2}gt^2,$$ (6.1)

the acceleration of gravity g is calculated with the formula

$$g = \frac{2s}{t^2} = (10.0 \pm 0.2)\frac{m}{s^2},$$

delivering a sufficient degree of accuracy for school instruction. Studying this phenomenon using smartphones' acceleration sensor should also be integrated in a more sophisticated instructional setting, connected with other phenomena [6–8] addressing the students' misconceptions which could be related to this concept [9]. Anyway, current research shows that using this method to study free fall and oscillation phenomena could at least increase curiosity and motivation of the students when they learn with smartphones or tablets as experimental tools [10].

References

1. Vogt P., Kuhn J., Gareis, S.: Beschleunigungssensoren von Smartphones: Möglichkeiten und Beispielexperimente zum Einsatz im Physikunterricht. Praxis der Naturwissenschaften Physik in der Schule **60**(7), 15–23 (Oct. 2011).
2. https://ogy.de/sparkvue
3. https://ogy.de/accelogger
4. Glück, M.: MEMS in der Mikrosystemtechnik: Aufbau, Wirkprinzipien, Herstellung und Praxiseinsatz Mikroelektromechanischer Schaltungen und Sensorsysteme. Vieweg+Teubner, Wiesbaden (2005)
5. Schnabel, P.: "Elektronik-Kompendium" (Keyword: MEMS- Micro-Electro-Mechanical Systems). www.elektronik-kompendium.de/sites/bau/1503041.htm
6. Kuhn, J.: Relevant information about using a mobile phone acceleration sensor in physics experiments. Am. J. Phys. **82**(2), 94 (2014)
7. Kuhn, J., Vogt, P., Müller, A.: Analyzing elevator oscillation with the smartphone acceleration sensors. Phys. Teach. **52**(1), 55–56 (2014)
8. Vogt, P., Kuhn, J.: Analyzing collision processes with the smartphone acceleration sensor. Phys. Teach. **52**(2), 118–119 (2014)
9. Hall, J.: iBlack box? Phys. Teach. **50**(5), 260 (2012)
10. Hochberg, K., Kuhn, J., Müller, A.: Using smartphones as experimental tools – Effects on interest, curiosity and learning in physics education. J. Sci. Educ. Technol. **27**(5), 385–403 (2018)

Going Nuts: Measuring Free-Fall Acceleration by Analyzing the Sound of Falling Metal Pieces

Florian Theilmann

7.1 Theorem II, Proposition II

The spaces described by a body falling from rest with a uniformly accelerated motion are to each other as the squares of the timeintervals employed in traversing these distances.

Galilei [1] presented the kinematics of a one-dimensional accelerated motion with ease and in terms of elegant geometry. Moreover, he believed, "Philosophy [i.e. physics] is written in this grand book—I mean the universe—which stands continually open to our gaze, but it cannot be understood unless one first learns to comprehend the language and interpret the characters in which it is written. It is written in the language of mathematics, and its characters are triangles, circles, and other geometrical figures, without which it is humanly impossible to understand a single word of it." [2]. In classroom practice, however, it can be difficult to reveal this mathematical heart of nature; free fall and other accelerated motions often get obscured by friction or other sources of errors. In this chapter we introduce a method of analyzing freefall motion indirectly by evaluating the noise of freely falling metal pieces. The method connects a deeper understanding of the mathematical structure of accelerated motion with the possibility to derive a numerical value for the free-fall acceleration g.

F. Theilmann (✉)
Department of Physics, University of Education Weingarten, Weingarten, Germany
e-mail: theilmann@ph-weingarten.de

7.2   The Experiment

The first corollary of the cited theorem is that for such an accelerated motion the distances traveled in equal intervals of time "will bear to one another the same ratio as the series of odd numbers, 1, 3, 5, 7." The proof is straightforward: The sequence of odd numbers adds up to the sequence of square numbers

$$1 = 1^2$$
$$1 + 3 = 4 = 2^2$$
$$1 + 3 + 5 = 9 = 3^2, \text{etc}$$

The general truth of this is realized by writing the difference of two consecutive integer squares, i.e., $(n + 1)^2 - n^2 = n^2 + 2n + 1 - n^2 = 2n + 1$, which is just the expression for the $(n + 1)$th odd number. Thus, as Galilei is claiming, "this [1 : 3 : 5 : 7 etc.] is the ratio of the differences of the squares of the natural numbers beginning with unity."

In physics classes, this corollary has often been used for a qualitative demonstration of Galilei's theorem: A number of small metal pieces (e.g., bolt nuts) are knotted to a cord with distances of appropriate ratio (Fig. 7.1). The cord is held in such a way

Fig. 7.1 A number of bolt nuts are attached to a cord. The distances between the nuts "bear to one another the same ratio as the series of odd numbers, 1, 3, 5, 7"

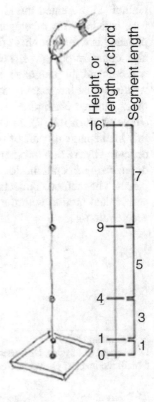

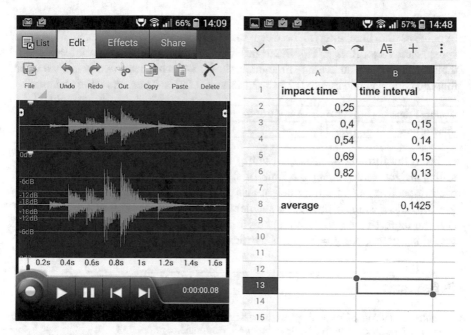

Fig. 7.2 Left: The sound profile from the sequence of beats produced by bolt nuts attached to a cord according to Fig. 7.1. Right: Spreadsheet for calculating the average time interval between the impacts of the bolt nuts on the floor

that the end with the shortest distance between the pieces just touches the floor (or, producing louder noise, a baking sheet lying on the floor). According to the corollary, dropping the cord results in a *periodic* sequence of percussions. In contrast, evenly spaced nuts produce an accelerating sequence of beats.

The procedure may be enhanced to a quantitative experiment for finding the free-fall acceleration g. As already shown in previous contributions (Chaps. 12 and 49) [3, 4], the drumming of the metal pieces used could be recorded with an appropriate app [5, 6]. For a subsequent sound analysis we edited the resulting sound file with a sound editor [6], cf. Figure 7.2, left. In the signal versus time representation all the consecutive beats reveal the same structure (with increasing amplitude corresponding to an increasing falling height), namely an instant onset of the signal with a subsequent exponential decrease.

7.3   Evaluating the Sound File

Timing the onset of the signal allows us to determine the respective moment of impact for the metal piece. This leads to the following table:

Falling height	0.105 m	0.39 m	0.89 m	1.59 m	2.46 m
Impact time	0.25 s	0.40 s	0.54 s	0.69 s	0.82 s

We may deduce the falling time of the metal pieces from the impact time; by hypothesis the time interval between two impacts should be constant. Using a spreadsheet (e.g., Ref. 7, Fig. 7.2 right) we may calculate the time intervals and their average. With an average time interval of 0.143 s between two beats, we infer that the cord was released at 0.25 s - 0.143 s = 0.107 s. Thus, the table becomes:

Falling height	0.105 m	0.39 m	0.89 m	1.59 m	2.46 m
Falling time	0.143 s	0.293 s	0.433 s	0.583 s	0.713 s

From the formula $h = 0.5gt^2$ (with g being the free-fall acceleration, h being falling height, and t being the time of falling), we may calculate $g = 2h/t^2$ for each respective pair of falling height and time. Then, the average from our data is $g = 9.57$ m/s^2, which is only 2.5% smaller than the literature value of $g = 9.81$ m/s^2. A linear regression between falling height and the values of 0.5 times the squares of the falling time gives $g = 9,64$ m/s^2, which is even better.

References

1. Galilei, G.: Dialogue Concerning Two New Sciences, translated by Henry Crew and Alfonso de Salvio. Macmillan, New York (1914). http://oll.libertyfund.org/titles/753
2. Galilei, G.: The Assayer, translated by Stillman Drake, pp. 237–238. University of Pennsylvania Press, Philadelphia (1957)
3. Schwarz, O., Vogt, P., Kuhn, J.: Acoustic measurements of bouncing balls and the determination of gravitational acceleration. Phys. Teach. 51, 312–313 (May 2013)
4. Kuhn, J., Vogt, P., Hirth, M.: Analyzing the acoustic beat with mobile devices. Phys. Teach. 52, 248–249 (April 2014)
5. https://ogy.de/smartrecorder
6. https://ogy.de/wavepadaudioeditor
7. https://ogy.de/googlesheets

The Atwood Machine Revisited Using Smartphones

8

Martín Monteiro, Cecilia Stari, Cecilia Cabeza, and Arturo C. Marti

The Atwood machine is a simple device used for centuries to demonstrate Newton's second law. It consists of two supports containing different masses joined by a string. Here we propose an experiment in which a smartphone is fixed to one support. With the aid of the built-in accelerometer of the smartphone, the vertical acceleration is registered. By redistributing the masses of the supports, a linear relationship between the mass difference and the vertical acceleration is obtained. In this experiment, the use of a smartphone contributes to enhance a classical demonstration.

8.1 Theory

The Atwood machine is a simple device invented in 1784 by the English mathematician George Atwood [1–3]. It consists of two objects of mass m_A and m_B, connected by an inextensible massless string over an ideal massless pulley [1]. Applying Newton's second law to each mass we obtain

$$m_A g - T = m_A a$$
$$T - m_B g = m_B a,$$

(8.1)

where g is the gravitational acceleration, T is the tension force, and a is the vertical acceleration. Eliminating the tension between these equations, we obtain

M. Monteiro (✉)
Universidad ORT Uruguay, Montevideo, Uruguay
e-mail: monteiro@ort.edu.uy

C. Stari · C. Cabeza · A. C. Marti
Universidad de la República, Montevideo, Uruguay
e-mail: cstari@fing.edu.uy; cecilia@fisica.edu.uy; marti@fisica.edu.uy

© The Author(s), under exclusive license to Springer Nature Switzerland AG 2022
J. Kuhn, P. Vogt (eds.), *Smartphones as Mobile Minilabs in Physics*,
https://doi.org/10.1007/978-3-030-94044-7_8

$$a = \frac{m_A - m_B}{m_A + m_B} g \qquad (8.2)$$

or, in terms of the mass difference Δm and the total mass M,

$$a = \frac{\Delta m}{M} g. \qquad (8.3)$$

As mentioned in the original Atwood's book, many possible experiments can be implemented using his machine [1]. One of the simplest possibilities, adopted here, is, keeping the total mass constant, to vary Δm by redistributing a set of weights. In this case, a linear relationship between the vertical acceleration and the mass difference is obtained.

Fig. 8.1 Experimental setup consisting of two supports (**a**) and (**b**), connected by a string and supported on two pulleys

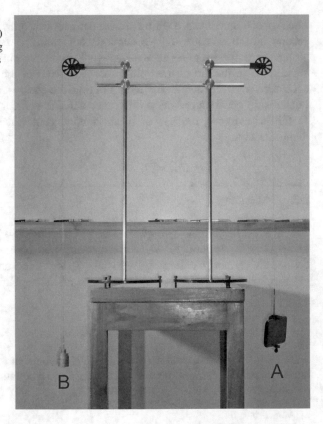

8.2   The Experiment

In our experimental setup, shown in Fig. 8.1, an Atwood machine was built using two pulleys. A smartphone is fixed on the right support (A), while on the left support (B), up to five weights can be placed. The smartphone, an LG G2, is kept fixed to the string as indicated in Fig. 8.2 using a clamp similar to those provided with tripods or monopods. In this experience, the smartphone is located in such a way that the only relevant axis is the x-axis, which coincides with the vertical direction.

Initially, the support A contains only the smartphone [the mass of the smartphone and support is $m_A = 191.2$ (2) g] while the support B holds five different weights [the mass of the support and the weights is $m_B = 190.8$ (2) g]. The total mass, $M = 382.0$ (4) g, is kept constant along the experiment. The system is released and the app Vernier Graphical Analysis [4], shown in Fig. 8.3, is used to record the acceleration values during the interval in which the support A is going downward and the support B upward. Of course, care should be taken to avoid hitting the smartphone against the floor.

Once the smartphone is stopped, the app is paused and a plot of the acceleration as a function of time exhibiting a region of constant acceleration or *plateau* is displayed

Fig. 8.2 Details of the support and the clamp

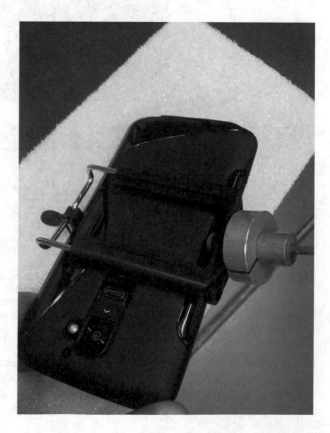

Fig. 8.3 Smartphone
mounted on the support
showing the Vernier app on
the screen

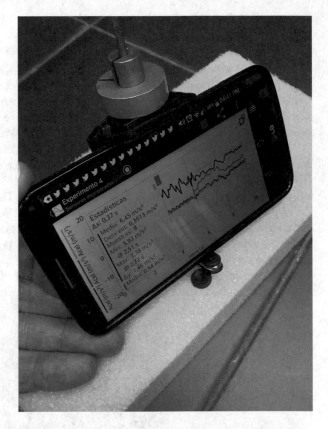

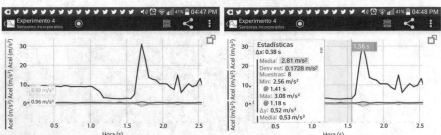

Fig. 8.4 Snapshots of the Vernier app showing the values registered by the acceleration sensor as a
function of time. The only relevant component here is the x, which corresponds to the vertical
acceleration. On the left panel, the plateau of constant acceleration can be appreciated. The
statistical values (mean and standard deviation) calculated for this interval are highlighted in the
right panel above

on the screen (see the left panel of Fig. 8.4). The vertical acceleration and its error are
obtained from the mean value and the standard deviation provided by the app
(Fig. 8.4). The acceleration measured includes the gravitational contribution, so it
is necessary to subtract it to obtain the real acceleration (Chap. 6) [5–7].

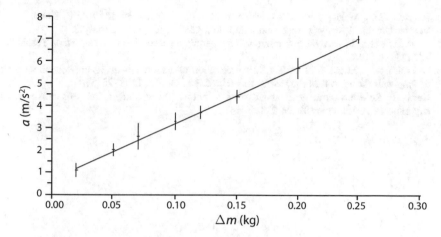

Fig. 8.5 Acceleration as a function of the mass difference: experimental points with error bars (black) and a linear regression (red line). Note that the yellow highlighted results from Fig. 8.4 are included here as the right-most data point after subtracting the acceleration due to gravity (9.8 m/s^2– 2.81 m/s^2 ≈ 7.0 m/s^2)

Subsequently, each of the weights is removed from the support B and placed on the support A. In this way, the mass of the system remains constant, and only the mass difference $\Delta m = m_A - m_B$ is varied. For each configuration the vertical acceleration is measured. Then, in Fig. 8.5, plotting the acceleration as a function of the mass difference, we obtain a straight line whose slope corresponds to g/M as indicated in Eq. (8.3).

8.3   Analysis and Conclusion

From the slope fitted in the linear regression, we obtain a value for the total mass of the system

$$M_{\exp} = \frac{g}{\text{slope}},$$

which results in $M_{\exp} = (387 \pm 20)$ g. This value is in considerable agreement with the value obtained by direct weighting, $M = (382.0 \pm 0.4)$ g. We conclude that, thanks to the aid of the accelerometer of a smartphone, it is possible to foster this demonstration and obtain a precise verification of Newton's second law.

References

1. Greenslade Jr., T.B.: Atwood's machine. Phys. Teach. **23**, 24 (Jan. 1985)
2. Siqueira, A., Almeida, A.d.C.S., Frejlich, J.: Máquina de Atwood. Rev. Bras. Ens. Fis. **21**, 95 (1999)

3. Johnson, G.O.: Making Atwood's machine 'work,'. Phys. Teach. **39**, 154 (March 2001)
4. Vernier Graphical Analysis App.: (n.d.). https://ogy.de/verniergraphicalanalysis
5. Vogt, P., Kuhn, J.: Analyzing free fall with a smartphone acceleration sensor. Phys. Teach. **50**, 182 (March 2012)
6. Monteiro, M., Cabeza, C., Martí, A.C.: Acceleration measurements using smartphone sensors: dealing with the equivalence principle. Rev. Bras. Ens. Fís. **37**, 1303 (2015)
7. Kuhn, J.: Relevant information about using a mobile phone acceleration sensor in physics experiments. Am. J. Phys. **82**, 94 (Feb. 2014)

Study of a Variable Mass Atwood's Machine Using a Smartphone

9

Dany López, Isidora Caprile, Fernando Corvacho, and Orfa Reyes

The Atwood machine was invented in 1784 by George Atwood [1] and this system has been widely studied both theoretically [2–3] and experimentally [4–5] over the years. Nowadays, it is commonplace that many experimental physics courses include both Atwood's machine and variable mass to introduce more complex concepts in physics. To study the dynamics of the masses that compose the variable Atwood's machine, laboratories typically use a smart pulley. Now, the first work that introduced a smartphone as data acquisition equipment to study the acceleration in the Atwood's machine was the one by M. Monteiro et al. [5] (Chap. 9). Since then, there has been no further information available on the usage of smartphones in variable mass systems. This prompted us to do a study of this kind of system by means of data obtained with a smartphone and to show the practicality of using smartphones in complex experimental situations.

9.1   Theory

The variable mass Atwood's machine consists of two variable masses $m_a(t)$ and $m_b(t)$. Without loss of generality, we considered that only $m_a(t)$ changes in time. Both masses are connected by an inextensible massless string over an ideal massless pulley. The acceleration equation of this system has been studied in depth by José Flores et al. [6]. Based on the results obtained there, the common acceleration is given by

D. López (✉)
Pontificia Universidad Católica de Chile, Facultad de Educación, Santiago, Chile
e-mail: dxlopez@uc.cl

I. Caprile · F. Corvacho · O. Reyes
Universidad de Chile, Facultad de Ciencias, Santiago, Chile
e-mail: isidora.caprile@ug.uchile.cl; fernando.corvacho@ug.uchile.cl; oreyes@u.uchile.cl

$$a = \left(\frac{m_a(t) - m_b}{m_a(t) + m_b}\right) g, \tag{9.1}$$

under the influence of an effective gravitational field g. As we can see, the integrability of Eq. (9.1) is largely dependent on $m_a(t)$. After some considerations that are well explained in Ref. [6] for $m_a(t) = c_0(t) + m_a(t = 0)$, the common speed is given by:

$$v(t) = v(t = 0) + \frac{gt}{\lambda} + \left(\frac{M_0 - \lambda m_{ab,0}}{\lambda^2 C_0}\right) g \ln\left(1 - \frac{C_0 \lambda t}{M_0}\right), \tag{9.2}$$

where

$$\lambda = 1 - \left(\frac{m_b}{M_0 - (1/2)m_a(t = 0)}\right),$$

and c_0 is the flow rate of sand when the system is not accelerating, $M_0(t) = m_b + m_a(t = 0)$ and $m_{ab,0} = m_a (t = 0) - m_b$.

Equation (9.2) can be compared directly with the results of our smartphone measurements.

9.2   The Experiment

Assembling

The experimental setup is shown to the left of Fig. 9.1. Two pulleys were located at the top. A negligible mass string was attached to it. A syringe (filled with sand) was added on the side of the string (position A in Fig. 9.1 left), which mimics our variable mass. On the other side of the string, a smartphone (iPhone 5) was attached (position B in Fig. 9.1 left). The "Vernier data acquisition" [7] program was used to collect the data over a certain period of time.

Experimental Procedure

The experimental procedure is similar to the one used in the recent work by M. Monteiro et al. [5]. However, it was necessary to check that the flow rate of sand had a constant decrease over time [6]. To do so, PASCO Force Sensor software was used (in Fig. 9.1 right, the experimental assembly is shown). This can be checked in the inset of Fig. 9.3, in which the flow of sand is clearly constant. Once we had considered all the factors mentioned above, the data could be collected. Initially the smartphone was placed at position B (the mass of the smartphone and support is $m_b = 0.15135 \pm 0.00001$ kg) while the syringe was placed at position A,

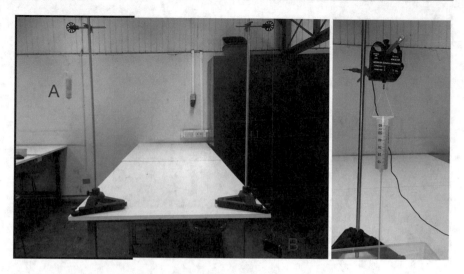

Fig. 9.1 Left: Experimental setup of the variable mass Atwood's machine. Right: Discharging flow of sand experimental setup. A PASCO Force Sensor was positioned at the top to collect the data

filled with sand (the initial mass of the syringe and sand is m_a $[t = 0] = 0.16170 \pm 0.00001$ kg).

Once the system is released, it is expected that each mass changes its velocity direction. In other words, as soon as the smartphone and syringe velocities go to zero due to the flow of sand, the smartphone velocity is expected to go from upwards to downwards. It's important to avoid hitting both the pulley against the syringe and the smartphone against the floor after observing the velocity change direction. Just before any of these happened, the system was stopped manually and the app was paused. The acceleration trend can be observed directly from the Vernier visualizer. An example is shown in Fig. 9.2. As expected for these variable mass systems, we observed a nonconstant acceleration over time (the area of interest is between ∼5 s to ∼11 s approximately in Fig. 9.2).

At the same time, the accelerations over time were also measured using the PASCO Smart Pulley software in order to compare the results obtained by the smartphone. The experiment was replicated several times, maintaining the initial proportions between the masses. All the other results were quite similar as the one shown in Fig. 9.2. To obtain the experimental velocity curve, the accelerations collected from the Vernier app (shown in the highlighted area in Fig. 9.2) were numerically integrated with respect to time by using the standard discrete integration formula.[1] After some minor adjustment of scale, the experimental velocity trend (collected using a smartphone) is shown in blue circles in Fig. 9.3 (this corresponds

---

[1] In order to integrate the experimental acceleration data numerically, we computed a discrete integration: $v_{n+1} = v_n + \left(\frac{a_{n+1}+a_n}{2}\right)(t_{n+1} - t_n), v_0 = 0.$

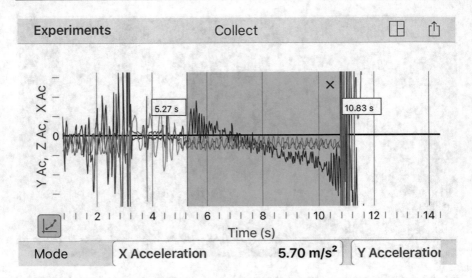

Fig. 9.2 Snapshot of the Vernier app. We observe three acceleration curves collected by the acceleration sensor. The yellow, blue, and red lines represent the x-, y-, and z-components, respectively. The only relevant component is the z. The interested area is highlighted in blue

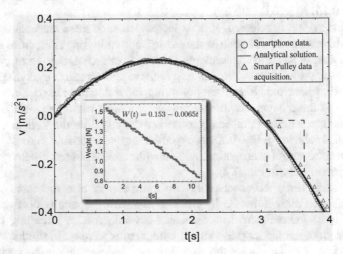

Fig. 9.3 Experimental result of the variable mass Atwood's machine. The circles represent the measured values of the velocity collected by a smartphone. The continuous line represents the analytical solution given by Eq. (9.2). The triangular dots represent the measured values of the velocity collected directly by using the Smart Pulley. The inset shows the linear flow rate of sand equal to 0.0065 N/s. The parameters of the system were $m_a (t = 0) = 0.16170 \pm 0.00001$ kg and $m_b = 0.15135 \pm 0.00001$ kg, $g = 9.78$ m/s^2

to one experimental realization only). Comparing this result with the analytical solution given by Eq. (9.2) (continuous line in Fig. 9.3), we observed a good agreement.

Finally, the data from the PASCO Smart Pulley software (triangular dots) are also plotted in Fig. 9.3. The results were qualitatively similar with respect to those obtained from the smartphone. Nevertheless, it is important to point out the fact that as soon as the masses changed their velocity directions, the Smart Pulley needed to change its rotational directions quickly, making it impossible to collect any data in that period of time. This inconvenience is highlighted in the dashed square area in Fig. 9.3. However, it is clear that a smartphone does not present any problem acquiring data when the system changes its velocity direction abruptly. This might be quite useful to laboratory courses in order to study more complex variable mass systems.

9.3 Conclusion

From Fig. 9.3 we can conclude that it is possible to study a variable mass system by using a smartphone. Indeed, a really good agreement between the experiment and theoretical prediction was observed. Moreover, using smartphones as data acquisition equipment instead of traditional laboratory equipment (such as a smart pulley) might be more adequate when systems present more complex dynamics, such as those faced when carrying out this experiment.

Despite the fact that this type of system is usually considered quite advanced to introduce in physics laboratory courses, it could be useful for students to get a deeper understanding about the physics behind variable mass systems. Moreover, introducing a smartphone as data acquisition equipment might lead students to think more carefully about the experimental data treatment and to question the meaning of the experimental results. Nonetheless, introducing both, variable mass Atwood's machine and smartphone, at the beginning might not be very useful and perhaps it is better to start with a more basic experiment, for instance those in Refs. [5, 6, 8], and to mention a few, before introducing this one.

Acknowledgments This work was carried out by D.L. in collaboration with two undergrad. Students, I.C. and F.C., as an undergraduate laboratory project in the Department of Physics of the Faculty of Sciences, U.C. We would like to acknowledge the valuable comments and constant support of Professor Mario Molina.

References

1. Greenslade Jr., T.B.: Atwood's machine. Phys. Teach. **23**, 24 (Jan. 1985)
2. McDermott, L.C., Shaffer, P.S., Somers, M.D.: Research as a guide for teaching introductory mechanics: an illustration in the context of the Atwood's machine. Am. J. Phys. **62**, 46–55 (Jan. 1994)

3. Tarnopolski, M.: On Atwood's machine with a nonzero mass string. Phys. Teach. **53**, 494–496 (Nov. 2015)
4. Wang, C.T.P.: The improved determination of acceleration in Atwood's machine. Am. J. Phys. **41**, 917–919 (July 1973)
5. Monteiro, M., Stari, C., Cabeza, C., Marti, A.C.: The Atwood machine revisited using smartphones. Phys. Teach. **53**, 373–374 (2015)
6. Flores, J., Solovey, G., Gil, S.: Flow of sand and a variable mass Atwood machine. Am. J. Phys. **71**, 715–720 (July 2003)
7. https://ogy.de/verniergraphicalanalysis
8. de Sousa, C.A.: Atwood's machine as a tool to introduce variable mass systems. Phys. Educ. **47**(2), 169–173 (2012)

Part III

Momentum and Collision

Analyzing Collision Processes with the Smartphone Acceleration Sensor

10

Patrik Vogt and Jochen Kuhn

Built-in acceleration sensors of smartphones can be used gainfully for quantitative experiments in school and university settings (see the overview in Ref. 1). The physical issues in that case are manifold and apply, for example, to free fall (Chap. 6) [2], radial acceleration (Chap. 16) [3], several pendula (Chaps. 29 and 32) [4, 5], or the exploitation of everyday contexts (Chap. 35) [6]. This chapter supplements these applications and presents an experiment to study elastic and inelastic collisions. In addition to the masses of the two impact partners, their velocities before and after the collision are of importance, and these velocities can be determined by numerical integration of the measured acceleration profile.

10.1 Theoretical Background and Execution of the Experiment

In the case of a central elastic impact, the paths of the centers of gravity of both impact partners lie on a straight line (Fig. 10.1). Frictional losses as well as changes of the potential energies are negligible compared to the kinetic energies of the impact partners. During the colliding process the system is isolated and without any impact of external forces [7].

Under the above-named requirements and in allowance of the conservations of momentum and of energy, it is possible to calculate the velocities of both impact

P. Vogt (✉)
Institute of Teacher Training (ILF) Mainz, Mainz, Germany
e-mail: vogt@ilf.bildung-rp.de

J. Kuhn
Ludwig-Maximilians-Universität München (LMU Munich), Faculty of Physics, Chair of Physics Education, Munich, Germany
e-mail: jochen.kuhn@lmu.de

Fig. 10.1 Straight collision

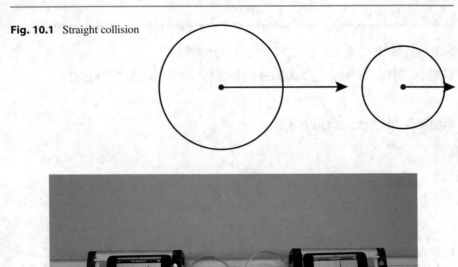

Fig. 10.2 Elastic collision

partners after the impact v'_1, v'_2 considering the starting velocity v_1, v_2 and the masses m_1, m_2 are given by:

$$v'_1 = \frac{(m_1 - m_2)v_1 + 2m_2v_2}{m_1 + m_2}$$

$$v'_2 = \frac{(m_2 - m_1)v_2 + 2m_1v_1}{m_1 + m_2}.$$

Analyzing these equations, it appears that if the masses of the impact partners are identical, they exchange their velocities and therefore their impulses and kinetic energies, which shall be tested by the first part of the experiment. For that purpose the experiment is arranged according to Fig. 10.2: Two carts of the same mass equipped with springs (a possible mass difference can be accommodated by attaching additional masses) are provided with either a smartphone or an iPod touch in such a way that their longitudinal axis (y-component of acceleration) is pointing in the direction of movement. The measuring frequency is set on the maximum value of 100 Hz (for iOS, e.g., the app SPARKvue [8], for Android, e.g., Accelogger [9]), the acceleration measurement started, and one of the carts pushed so that it collides with its partner.

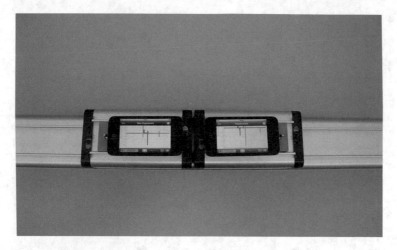

Fig. 10.3 Inelastic collision

In case of a centric, inelastic impact, some of the kinetic energy is lost due to friction and deformation. A special case is given for the inelastic impact where both impact partners leave the collision site with the same velocity v after the collision (a completely inelastic collision). Conservation of momentum gives us:

$$v = \frac{m_1 v_1 + m_2 v_2}{m_1 + m_2}.$$

In the case where the masses of both carts are alike and one cart is initially at rest before the inelastic collision, the impact partners keep moving according to $v = 0.5 \cdot v_1$. This is tested in the experiment shown in Fig. 10.3. Setup and realization take place analogously to the first part of the experiment, except that the springs are replaced by a hook-and-loop fastener that keeps the carts combined after the impact. Since the two carts have the same velocity after the inelastic impact, it would be sufficient for the experimental survey to use one smartphone or iPod touch, but to achieve constant cart masses it is easier to equip the second cart with a measurement device too.

10.2 Experiment Analysis

Central, Elastic Collision For the cart that is pushed by hand (cart 1), the $a(t)$-diagram shows a positive and a negative acceleration peak (Fig. 10.4). The first originates from pushing the static cart, which reaches (as shown by a numerical integration with the software Measure) a maximum velocity of $v_{1,\text{max}} \approx 0.53$ m/s. The degradation due to friction reduces this velocity to $v_1 \approx 0.48$ m/s immediately before the impact. The area between the second peak and the time axis provides the velocity reduction during the impact phase—the cart almost stops after the impact

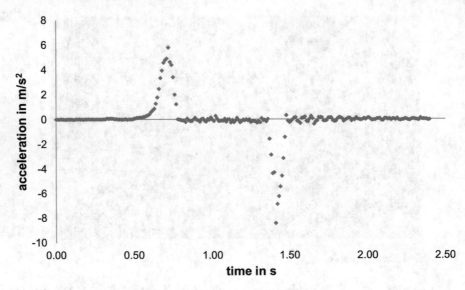

Fig. 10.4 Acceleration profile of cart 1 during the elastic impact

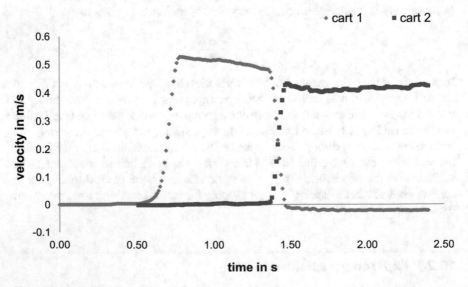

Fig. 10.5 Velocities during the elastic impact

and is just moving a little in the opposite direction ($v_1' \approx -0.02$ m/s). The acceleration measurement of cart 2 reveals that it picks up a velocity of 0.43 m/s during the impact, which is approximately v_1. Almost all the kinetic energy of cart 1 is transferred to the same weighted cart 2, showing approximate confirmation of conservation of momentum and of energy (Fig. 10.5).

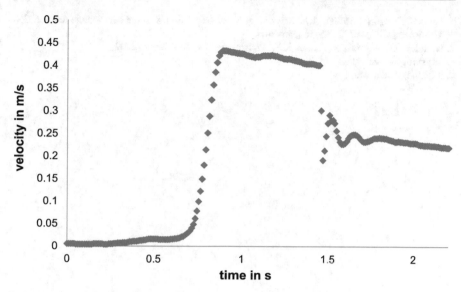

Fig. 10.6. Velocity of cart 1 during the inelastic impact

Central, Completely Inelastic Collision The velocity profile for cart 1 is shown graphically in Fig. 10.6. Cart 1 is accelerated to $v_{1,\text{max}} = 0.43$ m/s by an impulse and directly before striking cart 2 ($v_2 = 0$ m/s) it has the velocity $v_1 = 0.40$ m/s. After the impact, a short oscillation takes place at the hook- and-loop fastener before both carts move on collectively with 0.23 m/s. The velocity of cart 1 is thus approximately halved, which corresponds to the theoretical expectation. The small difference between theory and experiment results from the uncertainty of measurement and the test frequency of 100 Hz, which is relatively small for conducting collision experiments. After the impact the velocity of cart 2 is temporarily 0.19 m/s, which fits with the theoretical expectations very well, too.

References

1. Kuhn, J.: Relevant information about using a mobile phone acceleration sensor in physics experiments. Am. J. Phys. **82**(2), 94 (2014)
2. Vogt, P., Kuhn, J.: Analyzing free fall with a smartphone acceleration sensor. Phys. Teach. **50**, 182–183 (March 2012)
3. Vogt, P., Kuhn, J.: Analyzing radial acceleration with a smartphone acceleration sensor. Phys. Teach. **51**, 182–183 (March 2013)
4. Vogt, P., Kuhn, J.: Analyzing simple pendulum phenomena with a smartphone acceleration sensor. Phys. Teach. **50**, 439–440 (Oct. 2012)
5. Kuhn, J., Vogt, P.: Analyzing spring pendulum phenomena with a smartphone acceleration sensor. Phys. Teach. **50**, 504–505 (Nov. 2012)

6. Kuhn, J., Vogt, P., Müller, A.: Analyzing elevator oscillation with the smartphone acceleration sensor. Phys. Teach. **52**, 55–56 (Jan. 2014)
7. Halliday, D., Resnick, R., Krane, K.S.: Physics. Wiley, New York (2002)
8. https://ogy.de/sparkvue
9. https://ogy.de/accelogger

The Dynamics of the Magnetic Linear Accelerator Examined by Video Motion Analysis

11

Sebastian Becker-Genschow, Michael Thees, and Jochen Kuhn

A magnetic linear accelerator (or Gauss accelerator) is a device that uses the conversion of magnetic energy into kinetic energy to launch an object with high velocity. A simple experimental implementation consists of a line of steel spheres in which the first one is a permanent magnetic sphere. If another steel ball collides with the magnetic sphere from the left, the rightmost steel sphere is ejected at a much larger velocity than the impacting steel sphere. Several approaches have been published to determine the velocity of the ejected sphere, for example by using photogates [1, 2] or by measuring the impact position of the ejected sphere after falling from a table [3]. All of these approaches have in common that the measurement of the velocity is a static process. This contribution describes an approach in which video motion analysis on tablet computers [4, 5] is used to measure simultaneously the time course of the velocity of the impacting steel sphere as well as the ejected steel sphere, thus giving students insight into the dynamics of the processes that lead to a higher kinetic energy of the ejected sphere.

S. Becker-Genschow (✉)
Digitale Bildung, Department Didaktiken der Mathematik und der Naturwissenschaften,
Universität zu Köln, Cologne, Germany
e-mail: sebastian.becker-genschow@uni-koeln.de; sbeckerg@uni-koeln.de

M. Thees
Department of Physics/Physics Education Research Group, Technische Universität Kaiserslautern,
Kaiserslautern, Germany
e-mail: theesm@physik.uni-kl.de

J. Kuhn
Ludwig-Maximilians-Universität München (LMU Munich), Faculty of Physics, Chair of Physics
Education, Munich, Germany
e-mail: jochen.kuhn@lmu.de

11.1   Theoretical Background

We examine an arrangement of four spheres with equivalent volume consisting of one steel sphere and a line of one fixed permanent magnetic sphere with two other steel spheres. Friction forces should be neglected. If the single steel sphere is placed close to the magnetic sphere, the steel sphere is magnetized as soon as it enters the magnetic dipole field [6]. The interaction between the permanent magnetic dipole and the induced dipole results in an attractive force F_M leading to acceleration of the steel sphere (Fig. 11.1a). After the collision the kinetic energy is transferred through the line of spheres (Fig. 11.1b) to the rightmost sphere, which then is ejected at high velocity (Fig. 11.1c).

As the line of spheres is inhomogeneous because of the permanent magnetic sphere, the propagation of the energy through the line of spheres can be described as a soliton [7–9], which propagates through the line of spheres. Although the transmitted energy is smaller than the kinetic energy of the impacting steel sphere due to dissipation effects, the ejected sphere has a much larger velocity. The reason for this is that the increase of kinetic energy of the impacting sphere by conversion of the magnetic field energy exceeds the energy loss of the ejected sphere while it escapes the attractive magnetic field because of the greater distance from the permanent magnetic sphere. The kinetic energy of the ejected sphere can be estimated by [10]

$$E_{\text{kin}} \approx [\eta_0 - (m+1)0.024] \frac{\pi R^3 B_{\text{r}}^2}{36\mu_0}, \tag{11.1}$$

whereby R denotes the radius of the permanent magnetized sphere, B_r the residual flux density, η_0 the ratio of the kinetic energy of the ejected sphere and the kinetic

Fig. 11.1 Schematic of the different phases of the magnetic linear accelerator

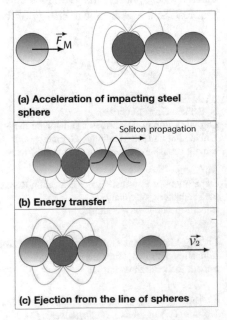

(a) Acceleration of impacting steel sphere

Soliton propagation

(b) Energy transfer

(c) Ejection from the line of spheres

energy of the impacting sphere in case of a line of steel spheres, μ_0 the vacuum permeability, and m the number of steel spheres right of the magnetic sphere (in our case $m = 2$). With Eq. (11.1) we can specify a formula for the velocity of the ejected steel sphere with a mass of m_S for our considered arrangement of four spheres.

$$v \approx \sqrt{2(\eta_0 - 3 \cdot 0.024)\frac{\pi R^3 B_r^2}{36\mu_0 m_s}}. \tag{11.2}$$

11.2   Experimental Setup

To reduce friction losses, an aluminum track is used to guide the spheres. The permanent magnetic sphere is attached to the track with putty to fix its position and prevent recoil. Two steel spheres are docked to the magnetic sphere to build a line. Another steel sphere is brought up to the magnetic sphere from the left (Fig. 11.2).

To measure the velocities, the distance between the steel sphere and the magnetic sphere is carefully reduced until the attraction causes the steel sphere to start speeding up. The motion is recorded with a tablet computer with a camera frequency rate of 120 frames per second (fps) perpendicular to the plane of motion from a sufficient distance (approximately 1 m). For recording and analyzing the process, we used an iPad Mini and the free application Viana [11].

11.3   Experimental Results

Based on the time-position data, the velocity of both spheres can be computed (see Fig. 11.3 left). The velocity of the ejected sphere can be extracted from the time-position data after the ejection by regression analysis with Vernier Graphical Analysis [12]. A linear regression of the data provided a value of 2.31 m/s with an RMSE value smaller than 0.001 (Fig. 11.3 right). Equation (11.2) can be used to calculate the theoretical value for the velocity of the ejected sphere. By inserting the following values, $m_S = 0.065$ kg, $R = 0.013$ m, $B_r = 1.22$ T [13], and $\eta_0 = 0.95$ [7], we calculated the theoretically expected value to be $v_{\text{theo}} = 2.48$ m/s. If we compare the

Fig. 11.2 Experimental setup

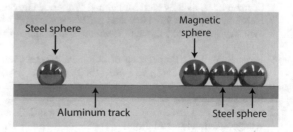

Steel sphere

Magnetic sphere

Aluminum track

Steel sphere

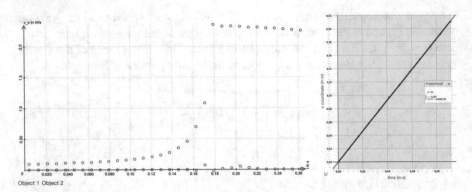

Fig. 11.3 Screenshots of time-velocity diagram (left, red: impacting sphere, blue: ejected sphere) and of least-squares proportional fit of time-position data (right)

calculated value with the experimentally determined value $v_{exp} = 2.31$ m/s, we find a relative deviation of 7.4%. A possible explanation for the smaller value of the experimentally determined value could be the neglect of both friction effects and the kinetic rotational energy of the incident sphere, which are not taken into account in the underlying theoretical model. In addition, the experimental setup could also be used to determine the magnetic flux density based on the measured velocity. Furthermore research shows that learning with mobile video analysis can also increase conceptual understanding [14, 15] while decreasing irrelevant cognitive effort and negative emotions [15, 16].

References

1. Rabchuk, J.A.: The Gauss rifle and magnetic energy. Phys. Teach. **41**, 158–161 (2003)
2. Andersson, Å., Karlsson, C.-J., Lane, H.: The Gaussian cannon. Emergent Sci. **1**, 6 (2017)
3. Kagan, D.: Energy and momentum in the Gauss accelerator. Phys. Teach. **42**, 24–26 (2004)
4. Becker, S., Klein, P., Kuhn, J.: Video analysis on tablet computers to investigate effects of air resistance. Phys. Teach. **54**, 440–441 (2016)
5. Klein, P., Gröber, S., Kuhn, J., Müller, A.: Video analysis of projectile motion using tablet computers as experimental tools. Phys. Educ. **49**(1), 37–40 (2014)
6. Jackson, J.D., Classical Electrodynamics, 3rd edn, pp. 231–233. Wiley, New York (1999), in German translation by Kurt Müller and Walter de Gruyter (2006)
7. Herrmann, F., Seitz, M.: How does the ball-chain work? Am. J. Phys. **50**, 977–981 (1982)
8. Sen, S., Manciu, M.: Discrete Hertzian chains and solitons. Physica A. **268**(3–4), 644–649 (1999)
9. Nesterenko, V.F.: Propagation of nonlinear compression pulses in granular media. J. Appl. Mech. Tech. Phys. **24**(5), 733–743 (1983)
10. Chemin, A., et al.: Magnetic cannon: the physics of the Gauss rifle. Am. J. Phys. **85**, 495 (2017)
11. https://ogy.de/Viana
12. https://ogy.de/VernierGraphicalAnalysis (for iOS) and https://ogy.de/VernierGraphical AnalysisAndroid (for Android)

13. https://www.supermagnete.de/data_sheet_K-26-C.pdf
14. Becker, S., Gößling, A., Klein, P., Kuhn, J.: Using mobile devices to enhance inquiry-based learning processes. Learn. Instr. **69**(2020), 101350 (2020)
15. Becker, S., Gößling, A., Klein, P., Kuhn, J.: Investigating dynamic visualizations of multiple representations using mobile video analysis in physics lessons: effects on emotion, cognitive load and conceptual understanding. Zeitschrift für Didaktik der Naturwissenschaften. **26**(1), 123–142 (2020)
16. Hochberg, K., Becker, S., Louis, M., Klein, P., Kuhn, J.: Using smartphones as experimental tools—A follow-up: cognitive effects by video analysis and reduction of cognitive load by multiple representations. J. Sci. Educ. Technol. **29**(2), 303–317 (2020)

Acoustic Measurements of Bouncing Balls and the Determination of Gravitational Acceleration

12

Oliver Schwarz, Patrik Vogt, and Jochen Kuhn

Interesting experiments can be performed and fundamental physical relationships can be explored with so-called Super Balls or bouncy balls. An example is the determination of gravity g in an experiment. The basic idea behind this was described by Pape [1] and Sprockhoff [2]: The initial and final heights and the complete duration of all the bounces are measured for a certain number of bounces by the ball. On the basis of this data, the acceleration of gravity can be approximately calculated if air drag on the ball is neglected. However, in practice, it becomes clear that measuring the height of the last bounce in the process is problematic. The person performing the experiment either has to make a good estimation of its height or film the bounce in front of a measuring stick. The method is based on the important assumption that each of the individual bounces of the ball loses the same percentage of mechanical energy; the coefficient of restitution k therefore remains the same.

12.1 Acoustic Data Measurement

Inspired by the research referred to above, our objective was to find an effective way of collecting data about a Super Ball's bouncing process in order to measure the accelaration of gravity, free fall, and the throw and the coefficients of restitution in an

O. Schwarz (✉)
Department of Physics–Didactics of Physics, University of Siegen, Siegen, Germany
e-mail: schwarz@physik.uni-siegen.de

P. Vogt
Institute of Teacher Training (ILF) Mainz, Mainz, Germany
e-mail: vogt@ilf.bildung-rp.de

J. Kuhn
Ludwig-Maximilians-Universität München (LMU Munich), Faculty of Physics, Chair of Physics Education, Munich, Germany
e-mail: jochen.kuhn@lmu.de

© The Author(s), under exclusive license to Springer Nature Switzerland AG 2022
J. Kuhn, P. Vogt (eds.), *Smartphones as Mobile Minilabs in Physics*,
https://doi.org/10.1007/978-3-030-94044-7_12

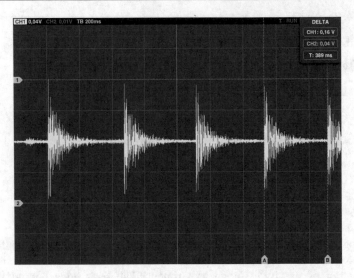

Fig. 12.1 Chronological sequence of the sound signals made by a bouncing ball

Fig. 12.2 Experiment setup
for the acoustic measurement
using an iPad

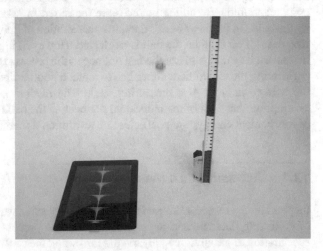

experiment. We found the use of an acoustic measure particularly effective, as
described in [3–5]. The sounds made by the impacts of the ball are recorded with
a microphone as voltage signals over a certain period of time. This produces a
chronological sequence for a Super Ball, with the sound made by the impacts
resulting in surprisingly sharp peaks, as can be seen in Fig. 12.1. These peaks can
be seen as time markers. The data were collected using an iPad equipped with the
Oscilloscope app [6], which can also be installed on an iPhone or iPod touch [7]. The
simple experiment setup can be seen in Fig. 12.2.

The person conducting the experiment should select the highest possible buffer
size (2000 ms), start the measurement, and then release the Super Ball onto a solid
hard, horizontal, smooth surface, e.g., a stone slab.

12.2 Determination of the Acceleration of Gravity

A vertically thrown ball rebounds upward between two impacts. The energy loss between two impacts could be represented by the coefficient of restitution $k = \frac{E_{\text{kin2}}}{E_{\text{kin1}}}$.

Taking fundamental equations into account [8], the coefficient of restitution could be calculated to

$$k = \frac{E_{\text{kin2}}}{E_{\text{kin1}}} = \frac{v_2^2}{v_1^2} = \frac{h_2}{h_1} = \frac{t_{\text{H2}}^2}{t_{\text{H1}}^2} = \frac{\Delta t_2^2}{\Delta t_1^2}. \tag{12.1}$$

It could be proved that the coefficient of restitution k remains constant, even for very different heights [9].

In order to determine g, the maximum height of the ball between two impacts has to be measured at least once during the bouncing process. It makes sense to select the initial height of the ball for this, which was 0.7 m for the experiment described below, as it is easily measured. The analysis was conducted as follows.

Assuming that the person performing the experiment has calculated the relative energy loss per impact k as described previously and has ascertained that the value remains constant from bounce to bounce, it is possible to determine the maximum height h_2 of the ball after its first impact with the floor. If h_1 designates the measured initial height, then the maximum height is given by:

$$h_2 = k \cdot h_1. \tag{12.2}$$

The free-fall time of the ball from its height h_2 until its impact is half of the time Δt between two impacts (Fig. 12.3). By taking this consideration, Eq. (12.2), and the distance-time law of free fall into account, g is obtained as follows:

$$g = \frac{2h_2}{(0.5 \cdot \Delta t)^2} = \frac{2kh_1}{(0.5 \cdot \Delta t)^2}. \tag{12.3}$$

Fig. 12.3 Determination of the critical sizes

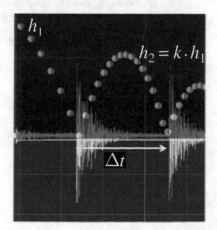

Table 12.1 Determination of gravitational acceleration on the basis of three impact times, with an initial height of 0.7 m each time

Impact times	Calculated value of g (m/s^2)
$t_1 = 0.248$ s, $t_2 = 0.955$ s, $t_3 = 1.617$ s	9.82
$t_1 = 0.201$ s, $t_2 = 0.898$ s, $t_3 = 1.549$ s	10.06
$t_1 = 0.129$ s, $t_2 = 0.830$ s, $t_3 = 1.479$ s	9.77

The results of the three measurements with the same initial height can be seen in Table 12.1. In conclusion, it is possible to measure the acceleration of gravity g and the relative energy loss of an impact using a good Super Ball with a single recording of the sound produced by the impacts. The experiment yields a result that is sufficiently accurate for the purposes of physics instruction.

Research shows that studying acoustic phenomena with mobile devices integrated in a more sophisticated instructional setting and combined with more everyday phenomena [10, 11] could also increase learning [12].

References

1. Pape, B.V.: Fallbeschleunigung mit einem hüpfenden Ball. Praxis der Naturwissenschaften – Physik in der Schule. **4**(49), 28–32 (2000)
2. Sprockhoff, G.: Physikalische Schulversuche, Mechanik. Oldenbourg Verlag, Munich/Düsseldorf (1961)
3. Schwarz, O., Vogt, P., Kuhn, J.: Acoustic measurements of bouncing balls and the determination of gravitational acceleration. Phys. Teach. **51**(5), 312–13 (2013)
4. White, J.A., Medina, A., Román, F.L., Velasco, S.: A measurement of g listening to falling balls. Phys. Teach. **45**, 175–177 (2007)
5. Aguiar, C.E., Laudares, F.: Listening to the coefficient of restitution and the gravitational acceleration of a bouncing ball. Am. J. Phys. **71**, 499–501 (2003)
6. https://ogy.de/oscilloscope
7. Alternatively, a commercial measuring system or a free sound editor (e.g., Audacity) can be used to make the acoustic recordings.
8. The kinetic energies E_{kin1} and E_{kin2} between two subsequent impacts behave like the squares of the impact velocities, and the rise and fall times t_H of the ball are given with Thereby we measure the time between two impacts, i.e., $\Delta t = 2t_H$.
9. Kuhn, J., Vogt, P.: Smartphones as experimental tools: different methods to determine the gravitational acceleration in classroom physics by using everyday devices. Eur. J. Phys. Educ. **4**(1), 16–27 (2013)
10. Vogt, P., Kuhn, J., Neuschwander, D.: Determining ball velocities with smartphones. Phys. Teach. **52**(6), 376–377 (2014)
11. Müller, A., Vogt, P., Kuhn, J., Müller, M.: Cracking knuckles – a smartphone inquiry on bioacoustics. Phys. Teach. **53**(5), 307–308 (2015)
12. Kuhn, J., Vogt, P.: Smartphone & Co. in physics education: effects of learning with new media experimental tools in acoustics. In: Schnotz, W., Kauertz, A., Ludwig, H., Müller, A., Pretsch, J. (eds.) Multidisciplinary Research on Teaching and Learning, pp. 253–269. Palgrave Macmillan, Basingstoke (2015)

Studying 3D Collisions with Smartphones 13

Vanda Pereira, Pablo Martín-Ramos, Pedro Pereira da Silva, and Manuela Ramos Silva

This chapter describes a conservation of momentum experiment using just smartphones and two beach balls, thus making the experimental study of this movement available to any classroom. For a more thorough analysis of the data, a computer can also be used. Experiments making use of smartphone sensors have been described before [1–12], contributing to an improved teaching of classical mechanics. In this experiment, we have made use of two smartphone cameras together with the VidAnalysis free app [13] to track the position of two balls colliding in air during a projectile motion (Fig. 13.1).

The app requires the setting of a length scale and the tracking of the ball position through screen touching, frame by frame, generating position-vs.-time and velocity-vs.-time graphs. The analysis has to be repeated for each ball. The data can easily be exported into a CSV file. Alternatively, the video can be transferred to a computer and converted into an image sequence (Fig. 13.2). The frames would subsequently be exported to a photo editor for retrieval of the x'-, y'-, and z'-coordinates of both balls, and the coordinate values would then be manipulated in a spreadsheet application like MS Excel or OpenOffice Calc to be transformed into real space coordinates, by using the height of the students as a scale factor (or that of any other object).

V. Pereira (✉) · P. P. da Silva · M. R. Silva
CFisUC, Department of Physics, FCTUC, Universidade de Coimbra, Coimbra, Portugal
e-mail: vanda.pereira@cpfs.mpg.de; psidonio@pollux.fis.uc.pt; manuela@pollux.fis.uc.pt

P. Martín-Ramos
EPSH, Universidad de Zaragoza, Huesca, Spain
e-mail: pmr@unizar.es

Fig. 13.1 Positioning of the smartphones and the students during the experiment

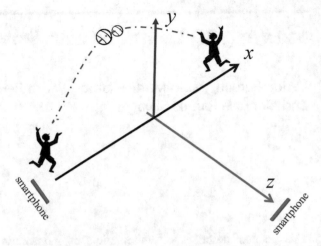

Fig. 13.2 Some of the frames taken during the flight of the balls

13.1 The Projectile Motion

Neglecting the resistance of air, a projectile is an object thrown near Earth's surface that will describe a curved path under the influence of gravity only. Before and after the collision, since there are no forces in the horizontal (x) and (z) axes, the acceleration of the motion is zero in those directions for both balls, their velocity is constant, and their position increases linearly with time. That is,

$$a_x = 0 \Rightarrow x(t) = x_0 + v_x t \tag{13.1}$$

$$a_z = 0 \Rightarrow z(t) = z_0 + v_z t \tag{13.2}$$

On the other hand, on the vertical axis (y) the gravity force acts on the object, causing a constant acceleration that points downwards,

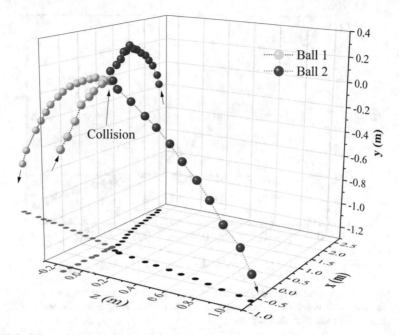

Fig. 13.3 3D graph of the position of the balls before and after collision

$$a_y = -g \Rightarrow v_y = v_{0y} - gt \Rightarrow y(t) = y_0 + v_{0y}t - 1/2gt^2. \qquad (13.3)$$

Therefore, before and after the collision, each of the balls describes a parabolic trajectory (Fig. 13.3).

13.2   The Conservation of Momentum

Neglecting the resistance of the air and apart from gravity, there are not any exterior forces acting on the system of the balls during collision. This means that conservation of the system momentum along the x- and z-axes has to be observed (Eqs. 13.4 and 13.5).

$$\begin{aligned}(m_1 v_{1x} + m_2 v_{2x})_{\text{(before collision)}} \\ = (m_1 v_{1x} + m_2 v_{2x})_{\text{(after collision)}}\end{aligned} \qquad (13.4)$$

and

$$\begin{aligned}(m_1 v_{1z} + m_2 v_{2z})_{\text{(before collision)}} \\ = (m_1 v_{1z} + m_2 v_{2z})_{\text{(after collision)}}\end{aligned} \qquad (13.5)$$

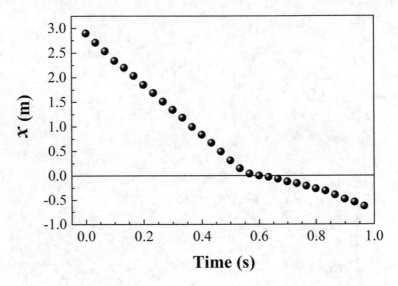

Fig. 13.4 Plot of x as a function of t for one of the balls. The origin of the coordinate axes was chosen to be the collision point. The slope was retrieved from fitting a straight line to the points immediately before and after the collision

Table 13.1 Velocities and momenta before and after the collision

		Ball 1	Ball 2	Momentum of the system $(kg \cdot m \cdot s^{-1})$
Velocity $(m \cdot s^{-1})$	Before collision	$v_{1x} = 5.04$	$v_{2x} = -5.01$	$p_x = 0.09$
		$v_{1z} = 0.02$	$v_{2z} = 0.60$	$p_z = 0.08$
	After collision	$v_{1x} = 1.69$	$v_{2x} = -1.38$	$p_x = 0.07$
		$v_{1z} = -2.03$	$v_{2z} = 3.35$	$p_z = 0.14$
Variation of momentum $(kg \cdot m \cdot s^{-1})$				$\Delta p_x = -0.03$ $\Delta p_z = +0.06$

The values of the pre- and post-collision velocities are very easy to retrieve from the slope of the $x(t)$ or $z(t)$ graphs (Fig. 13.4). The values for the pre- and post-collision velocities retrieved from the fitting of the $x(t)$ and $z(t)$ graphs for both balls are summarized in Table 13.1. Using the mass of the balls ($m_1 = 0.150$ kg and $m_2 = 0.132$ kg), it is possible to calculate the linear momentum of the system and confirm its conservation. The conservation of momentum was therefore easily demonstrated.

It is worth noting that, although the entire experiment can be done on a smartphone, it may be easier to process in VidAnalysis if a tablet is used instead, provided that the larger screen size improves the precision when the length of a known distance is specified and so as to define the center of the ball in each of the frames. It is also important to find a trade-off in terms of the distance at which the smartphones are from the students (larger distances lead to smaller distortion in the images, but ball positions in the frame-by-frame analysis become more difficult to follow in a precise manner).

This experiment can easily be changed to further motivate advanced students: using balls with very different masses instead of similar balls, using very elastic balls to probe the conservation of energy, trying a collision involving three balls, retrieving the acceleration of gravity through fitting of the y-coordinates with a polynomial expression (e.g., in this collision, the graph of $y(t)$ for ball 2, before collision, was fitted with the expression $y(t) = -4.92t^2 + 3.73t - 0.48$, which yielded 9.84 m/s^2 for gravity). This sort of experiment will engage both teachers and students since it overcomes the lack of resources, demands no time for pre-lab experiments assembling, and it uses their own gadgets.

Acknowledgments CFisUC gratefully acknowledges funding from FCT Portugal through grant UID/FIS/04564/2016. P.M-R would like to thank Santander Universidades for its financial support through the "Becas Iberoamérica Jóvenes Profesores e Investigadores, España 2016" scholarship program.

References

1. Chevrier, J., Madani, L., Ledenmat, S., Bsiesy, A.: Teaching classical mechanics using smartphones. Phys. Teach. **51**, 376 (2013)
2. Hall, J.: More smartphone acceleration. Phys. Teach. **51**, 6 (2013)
3. Mau, S., Insulla, F., Pickens, E.E., Ding, Z., Dudley, S.C.: Locating a smartphone's accelerometer. Phys. Teach. **54**, 246–247 (2016)
4. Monteiro, M., Stari, C., Cabeza, C., Marti, A.C.: The Atwood machine revisited using smartphones. Phys. Teach. **53**, 373–374 (2015)
5. Shakur, A., Kraft, J.: Measurement of Coriolis acceleration with a smartphone. Phys. Teach. **54**, 288–290 (2016)
6. Tornaría, F., Monteiro, M., Marti, A.C.: Understanding coffee spills using a smartphone. Phys. Teach. **52**, 502–503 (2014)
7. Vogt, P., Kuhn, J.: Analyzing simple pendulum phenomena with a smartphone acceleration sensor. Phys. Teach. **50**, 439 (2012)
8. Vogt, P., Kuhn, J., Müller, S.: Experiments using cell phones in physics classroom education: the computer-aided g determination. Phys. Teach. **49**, 383–384 (2011)
9. Becker, S., Klein, P., Kuhn, J.: Video analysis on tablet computers to investigate effects of air resistance. Phys. Teach. **54**, 440–441 (2016)
10. Gröber, S., Klein, P., Kuhn, J.: Video-based problems in introductory mechanics physics courses. Eur. J. Phys. **35**(5), 055019 (2014)
11. Klein, P., Kuhn, J., Müller, A., Gröber, S.: In: Schnotz, W., Kauertz, A., Ludwig, H., Müller, A., Pretsch, J. (eds.) Multidisciplinary Research on Teaching and Learning, pp. 270–288. Palgrave Macmillan UK, London (2015)
12. Klein, P., Gröber, S., Kuhn, J., Müller, A.: Video analysis of projectile motion using tablet computers as experimental tools. Phys. Educ. **49**(1), 37–40 (2014)
13. https://ogy.de/vidanalysis

Part IV
Rotation

Measuring Average Angular Velocity with a Smartphone Magnetic Field Sensor

14

Unofre Pili and Renante Violanda

The angular velocity of a spinning object is, by standard, measured using a device called a tachometer. However, by directly using it in a classroom setting, the activity is likely to appear as less instructive and less engaging. Indeed, some alternative classroom-suitable methods for measuring angular velocity have been presented [1, 2]. In this chapter we present a further alternative that is smartphone-based, making use of the real-time magnetic field (simply called B-field in what follows) data gathering capability of the B-field sensor of the smartphone device as the timer for measuring average rotational period and average angular velocity. The in-built B-field sensor in smartphones has already found a number of uses in undergraduate experimental physics. For instance, in elementary electrodynamics, it has been used to explore the well-known Bio-Savart law (Chap. 62) [3] and in a measurement of the permeability of air [4].

14.1 Theory

For any uniformly rotating object, its average angular velocity ω_{ave} is given by the relation [5]

$$\omega_{\text{ave}} = \frac{\Delta\theta}{\Delta t}.$$ (14.1)

Given the average period T_{ave}, Eq. (14.1) can be written as

U. Pili (✉) · R. Violanda
University of San Carlos, Cebu, Philippines
e-mail: ubpili@usc.edu.ph; rrviolanda@usc.edu.ph

© The Author(s), under exclusive license to Springer Nature Switzerland AG 2022
J. Kuhn, P. Vogt (eds.), *Smartphones as Mobile Minilabs in Physics*,
https://doi.org/10.1007/978-3-030-94044-7_14

$$\omega_{\text{ave}} = \frac{2\pi}{T_{\text{ave}}}. \tag{14.2}$$

Equation (14.2), with known value of the average period, easily gives the average angular velocity.

14.2   Experiment

Our experimental setup, with the rotational period being the parameter to be directly measured, is made up of a smartphone installed with an Android application called Physics Toolbox Sensor Suites [6], a computer with MS Excel, a small permanent magnet, and a non-precessing electric fan but with a protective slow-spinning grill. The fan, a China-made, has the following specifications—common name: Rota-aire; brand name: 3D New Generation; model: BF35EF; diameter: 35 cm. Instead of the precession of the traditional fan, the spinning grill allows for the distribution of air over a wider area, and measuring this slow and uniform angular velocity of the grill (not the measurement of the high angular velocity of the rotating blades) is the primary goal of this experiment. Our setup (sans the computer) is shown in Fig. 14.1. In this experiment, at least, we have to settle for measuring the slow rotational velocity of the grill rather than that of the fast-spinning blades of the fan because the sampling rate of the smartphone device was limited to only every 0.10 s, which means that the device was likely to return an erroneous measurement of the rotational period of an object that rotates with a period of less than 0.10 s. This can be the case for the fan blades rotating at a rate of 50 or 60 Hz. Besides, a parallel video-based analysis of high-speed rotational motion would not be as accessible.

Fig. 14.1 A non-precessing electric fan that uses a slow-spinning grill instead in order to distribute the moving air. A permanent magnet is glued (shown by the red arrow) on the edge of the grill

Fig. 14.2 Plot of total *B*-field as a function of time. Very slight outward or inward movements of the grill (or of the phone) caused the differences in amplitudes, but these were irrelevant since the device was simply used as the timer

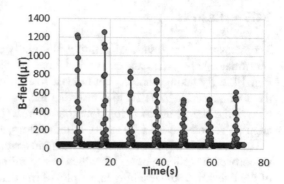

Starting off the data gathering, the magnetic field sensor of the device was launched via the Android application and, by holding or hanging the smartphone with its front side positioned closer to the spinning grill, numeric and graphical real-time measurements of the *B*-field of the magnet were registered. Directly saved as an MS Excel file in the device, the raw data were subsequently exported to the computer via email or Bluetooth for further processing, also in MS Excel. A computer-generated graph of the *B*-field data as a function of time is shown in Fig. 14.2.

The application allows for the extraction of the *x*-,*y*-, and *z*-components of the *B*-field as well as its total. All of these time-based *B*-field values can be utilized, giving the same period, but we chose to utilize the total *B*-field. The periodic time-varying total *B*-field, observed in Fig. 14.2, is due to the fact that its magnitude at an observation point varies with its distance from the source. Specifically, at a field point located on a given axis, the *B*-field due to a small permanent magnet—for a small cylindrical magnet, like the one used in this experiment, whose length is insignificant in comparison to the distance between its location and the field point—is inversely proportional to the cube of the distance [7]. The experimental setup does not exactly conform though to this specific *B*-field-distance relationship because the changes in location (or distance) of the magnet relative to the sensor (or smartphone) are angular in nature. Instead, the series of peaks seen in Fig. 14.2 can be explained in this manner. Each instant the magnet reaches a certain point in the smartphone (one closest to the sensor if not its exact location), the *B*-field peaks up, all the while becoming zero while the magnet rotates farther away from the smartphone. The pattern shown, however, is still a manifestation of the fact that the closer the field point is to the source, the stronger the *B*-field is at that point and vice versa. It can also be deduced that the *B*-field measurements (Fig. 14.2) are that of the magnet only, and there was no contributing *B*-field due to eddy currents since the blades and the grill are entirely made of plastic.

14.3   Results

The measure of the rotational period was directly obtained by subtracting the time coordinates, read via data cursor function, of any pair of successive peaks in Fig. 14.2. By taking into account at least five pairs of successive peaks, we found the average period to be 10.36 s. Inserting this value in Eq. (14.2) and subsequently converting the result to revolutions per minute (rpm), we obtained the average angular velocity to be 5.792 rpm. Alternatively, by plotting the angular position [8] (over a full rotation) of the magnet vs. time (over one period), the average angular velocity of the grill can be found, which is, according to Eq. (14.1), equal to the slope of the linear fit to the resulting linear plot of the data points (Fig. 14.3). It is equal to 5.77 rpm (also after conversion to rpm). In order to present further the uniformity of the average angular velocity [9], we have plotted its value/s over one period or one full rotation. Figure 14.4 presents such velocity against time plot. There was no accepted value indicated for the angular velocity of the spinning grill of the electric fan, but a parallel video analysis gave $\omega_{ave} = 5.8$ rpm, which is a good match in comparison to the result obtained using the method presented. In doing the video analysis, we played the movie, frame by frame, in Windows Media Player, allowing us to monitor the total time it took for the spinning grill (or magnet) to cover a full rotation.

14.4   Conclusions

The experiment presented has shown that the real time-measuring B-field sensor of a smartphone device is reliable as a timer for the measurement of rotational period, thus the accuracy in the measurement of angular velocity, albeit of a slow-spinning

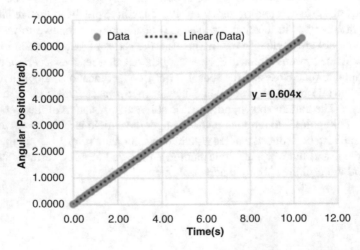

Fig. 14.3 Plot of angular position of the magnet vs. time over a full rotation along with a linear fit to the data points

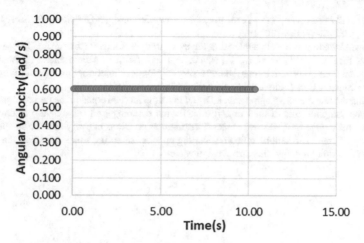

Fig. 14.4 Plot showing the uniform average angular velocity $\omega_{ave} = 0.604$ rad/s $= 5.77$ rpm over a full rotation of the grill (or magnet) against time

object like the one used in this experiment. This was done purposely so that a parallel video analysis can easily be performed. Indeed, our experiment appears to be limited only in measuring the angular velocity of slow-rotating objects, one in which the rotational period is at least equal to or greater than 0.10 s, the time resolution of the B-field sensor of the smartphone device. Nonetheless the setup is relatively low cost, considering that smartphones and computers are household articles nowadays; the slow-rotating grill of an electric fan can easily be replaced by other slow-rotating objects. Moreover, should an adoption of this experiment be considered as a classroom or laboratory activity, the students will be able to acquire a certain skill level on data mining. This is apart from a relatively enriched hands-on exposure to the concept of angular velocity.

The activity will also afford the students a lesson (before attending a regular class on elementary electromagnetism) on the distance dependence of the strength of magnetic field.

Acknowledgment We would like to thank our colleagues in the Physics Department, University of San Carlos, for their insightful comments. Our expression of gratitude also goes to the anonymous reviewer whose comments and suggestions made the manuscript a lot better.

References

1. Misra, R.M.: A simple method to measure the angular speed of a spinning object. Phys. Teach. **46**, 97 (2008)
2. "How to measure RPM of motor," YouTube, https://www.youtube.com/watch?v=PZby_Y3iuuk
3. Monteiro, M., Stari, C., Cabeza, C., Marti, A.C.: Magnetic field 'flyby' measurement using a smartphone's magnetometer and accelerometer simultaneously. Phys. Teach. **55**, 580 (2017)
4. Lara, V., Amaral, D., Faria, D., Vieira, L.: Demonstrations of magnetic phenomena: measuring the air permeability using tablets. Phys. Educ. **49**(6), 658 (2014)

5. Young, H.D., Freedman, R.A., Ford, A.L.: Sears and Zemansky's University Physics, 13th edn, p. 279. Addison-Wesley (2012)
6. We downloaded the Android application Physics Toolbox Suites for free from Google Play
7. Arribas, E., Escobar, I., Suarez, C.P., Najera, A., Beléndez, A.: Measurement of the magnetic field of small magnets with a smartphone: a very economical laboratory practice for introductory physics courses. Eur. J. Phys. **36**(6), 065002 (2015)
8. Knowing the fact that the magnet travels 2π rad in a time interval equal to the period together with the given sampling rate of total time it took for the spinning grill (or magnet) to cover a full rotation. the device of 0.10 s, we obtained the angular positions at every specific sampling time
9. Division of angular positions by corresponding specific sampling times generates the uniform values of angular velocity, thus the average velocity

Visualizing Acceleration with AccelVisu2 15

Thomas Wilhelm, Jan-Philipp Burde, and Stephan Lück

Acceleration is a physical quantity that is difficult to understand and hence its complexity is often erroneously simplified. Many students think of acceleration as equivalent to velocity $\vec{a} \sim \vec{v}$. For others, acceleration is a scalar quantity, which describes the change in speed $\Delta|\vec{v}|$ or $\Delta|\vec{v}|/\Delta t$ (as opposed to the change in velocity). The main difficulty with the concept of acceleration therefore lies in developing a correct understanding of its direction [1–4]. The free iOS app "AccelVisu2" [5] supports students in acquiring a correct conception of acceleration by showing acceleration arrows directly on the screen of an iPhone, iPad or iPod touch, e.g. when moving it across a table.

15.1 Theoretical Background

In terms of determining the direction of acceleration, three cases can be distinguished: when gaining speed, the acceleration is in the direction of movement, when slowing down, the acceleration is in the opposite direction to the direction of movement, and when driving around a bend at a constant speed, it is perpendicular to the direction of movement. Generally, the acceleration can be broken down into a

T. Wilhelm (✉)
Department of Physics Education Research, Goethe University Frankfurt, Frankfurt am Main, Germany
e-mail: wilhelm@physik.uni-frankfurt.de

J.-P. Burde
Physics Education Research Group, University of Tübingen, Tübingen, Germany
e-mail: Jan-Philipp.Burde@uni-tuebingen.de

S. Lück
Department of Physics, University of Würzburg, Würzburg, Germany
e-mail: slueck@physik.uni-wuerzburg.de

tangential component indicating the change of speed, and a radial component indicating the change of direction.

The acceleration of the iPhone is detected component-wise by the built-in accelerometer and displayed graphically as a vector in the coordinate system of the app. The acceleration sensor in the device measures the deflection of a test object, which in turn is a measure of the force acting on the seismic mass. In general, this force can be resolved into two components: the inertial force resulting from the acceleration of the device and the force of gravity. If preferred, it is possible to ignore the latter one in AccelVisu2 in order to only measure the acceleration of the device itself as explained in the next section.

15.2 The App AccelVisu2

Once the user starts the app AccelVisu2, a coordinate system is displayed (see Fig. 15.1). Here, the measured acceleration in the xy-plane is shown as a blue vector arrow. In addition, the x- and y-component of the measured acceleration are

Fig. 15.1 Screenshot of "AccelVisu2"

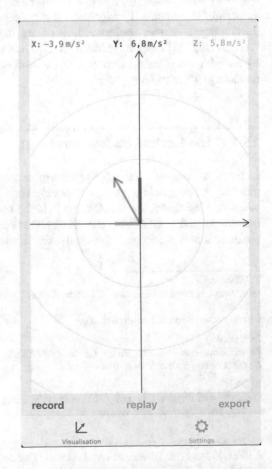

displayed in the form of a green and red line in the direction of the corresponding axis. At the top, the momentary acceleration values of all three acceleration components are shown (in m/s^2). The circular rings indicate multiples of the gravitational acceleration g. The coordinate system can be moved by "sliding" the finger over it. It is also possible to zoom in and out with the familiar "two finger pinch-gesture". A "double tap" on the screen restores the original view.

Below the diagram are the three buttons "record", "replay", and "export". After starting the app, the "record" mode is active. In this mode the acceleration data is stored internally with an adjustable data rate and length (Fig. 15.2). After the recording is stopped by tapping the button again or the chosen recording time has

Fig. 15.2 The setting menu in "AccelVisu2"

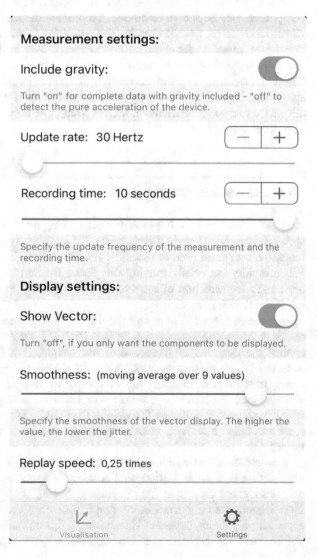

passed, the recorded acceleration data with the corresponding visualization can be repeated by tapping "replay". By tapping on "export" an e-mail with a table of the recorded data can be sent. By selecting the "Settings" tab (at the very bottom), the settings screen opens. As shown in Fig. 15.2, this screen is divided into two sections: "Measurement settings" and "Display settings".

If the toggle switch "Include gravity" is switched off, only the actual acceleration is measured (i.e. internally the gravitational acceleration is subtracted vectorially at any time). The "Recording rate" and "Recording time" sliders can be used to set the recording parameters as required. The recording rate is limited by the specifications of the respective device (max. 100 Hz), the longest recording time is 600 s.

Using the "Show Vector" button, the acceleration vector resulting from the components can be shown or hidden. To prevent the measured values and thus the visualization arrow from fluctuating too much due to the high measuring frequency, the "Smoothness" slider can be used to average over several values. As a result of this adjustable time averaging, outliers are smoothed out, which is particularly helpful for freehand experiments or for a car or carousel ride. The last slider "Replay speed" can be used to change the speed of playback of a recorded movement. This allows the user to create a slow motion or time-lapse effect to better track very fast or very slow movements.

15.3 Ideas for Experiments

First, let's consider two simple scenarios: picking up speed from a rest position and slowing down. As slightly tilting the device can already greatly affect the measurement, the user is advised to place the iPhone or iPod Touch on a horizontal table. When rapidly but briefly moving the phone forward, the acceleration vector first appears in the direction of movement while gaining speed and then counter to it while decelerating the device.

It is better, however, to mount the iPhone on a wagon that is pulled by a weight over a deflection pulley as shown in Fig. 15.3. If the user initially pushes the wagon against the direction of the force of the weight, it will continuously slow to a standstill before picking up speed again in the opposite direction. The app makes it easy to see that the acceleration vector neither changes in length nor direction in either phase of the movement, since the vector arrow always points in the direction in which the wagon is pulled by the weight. Contrary to what most students believe, the same acceleration arrow is clearly visible even at the turning point of the wagon's movement.

Another excellent application of the software is to visualize the direction of acceleration in circular motion with acceleration vectors. For example, the iPhone can be mounted on a rotatable experimental setup (Fig. 15.4). As long as there is little friction, the only acceleration that can be observed is the constant centripetal acceleration pointing inwards (when ensuring that one component of the acceleration will be the radial acceleration and the other the tangential acceleration). Furthermore,

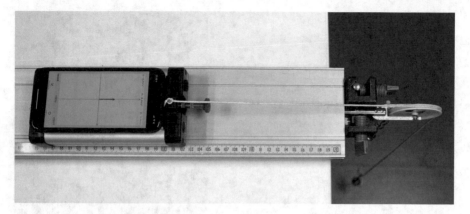

Fig. 15.3 Accelerated motion on a track

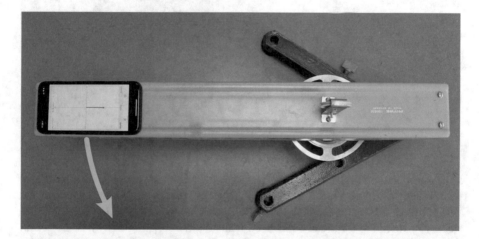

Fig. 15.4 Acceleration of a rotating iPhone

one can clearly see that the length of the acceleration arrow is dependent on the rotational speed of the iPhone.

Alternatively, a volunteer could sit on a well-oiled swivel chair and then be put into rotation by another individual. When holding the smartphone at arm's length, the volunteer can see the screen while rotating around her own axis. When conducting this experiment, it is important to switch off the option "include gravity" as it is nearly impossible to keep the device in a perfectly horizontal position all the time. Moreover, the visualization of the acceleration vector can be optimized via the "Softness" setting in order to smooth out vibrations from holding the device with your bare hands.

It is also possible to analyze the direction of acceleration in simple or rod pendulums by attaching the device to a well-suspended pendulum rod while making

Fig. 15.5 Top view of a pendulum standing on a table with two coil springs attached to a wall on the left. In this setup, the restoring force is caused by the coil springs instead of gravity

sure that the screen is always visible. Since the iPhone's position in a vertically suspended-pendulum would constantly change in relation to the gravity field, the pendulum should be positioned horizontally on a table as shown in Fig. 15.5. The necessary restoring force can be realized using a torsion spring or by a combination of two coil springs and pulleys. With the help of the AccelVisu2 app, it is then possible to see how the direction of the acceleration arrow changes during the movement of the pendulum. The acceleration arrow is generally inclined in the coordinate system, since both the speed (tangential acceleration component) and the direction (radial acceleration component) change with the movement. However, there are two interesting exceptions: In the reversal points, the acceleration arrow is only directed tangentially, and at the zero crossing, it is only directed radially. We would advise that users record a video of the smartphone's movement, since this will allow for analyzing the movement step-by-step at a later point, such as in the classroom. Generally, when conducting measurements in the classroom, we recommend using a projector which is wirelessly connected to an iOS device via "AirPlay" (e.g. with an AppleTV or a computer).

Another possible application is to observe the acceleration arrow while driving. Ideally, this should be done on as flat a road as possible. Furthermore, the "Softness" level (moving average) should be increased using the slider. Nonetheless, it is also advisable to use a bump-free road, otherwise the jerking of the vehicle will cause a strong wobble of the acceleration vector which may also make it impossible to identify the car's long-term accelerations, e.g. in curves. However, if the iPhone is in a fixed position, e.g. by attaching it to the center console in the car, the passenger can observe how the acceleration arrow changes when picking up speed, slowing down or driving around a bend. The main advantage to this application is that students can link the acceleration they experience with the acceleration vector they observe in order to develop an accurate conception of the physical quantity "acceleration". Similarly, the recorded data of a car ride (max. 10 min) can be shown at a later time (e.g. in a classroom) using the "playback" button. Here it has proven particularly helpful to record the car ride with a video camera (using a second smartphone) and then play the real video combined with a screen recording of AccelVisu2 to illustrate how the direction of acceleration changes during the ride. To combine the two videos, any simple video editing program can be used.

References

1. Flores, S., Kanim, S., Kautz, C.: Student use of vectors in introductory mechanics. Am. J. Phys. **72**(4), 460–468 (2004)
2. Hestenes, D., Wells, M.: A Mechanics baseline test. Phys. Teach. **30**, 159–166 (1992)
3. Labudde, P., Reif, F., Quinn, L.: Facilitation of scientific concept learning by interpretation procedures and diagnosis. Int. J. Sci. Educ. **10**, 81–98 (1988)
4. Wilhelm, T.: Vektorverständnis und vektorielles Kinematikverständnis von Studienanfängern. In: Nordmeier, V., Oberländer, A., Grötzebauch, H. (eds.) Didaktik der Physik – Regensburg 2007. Lehmanns Media – LOB.de, Berlin (2007)
5. https://ogy.de/accelvisu2

Analyzing Radial Acceleration with a Smartphone Acceleration Sensor

16

Patrik Vogt and Jochen Kuhn

This chapter continues the sequence of experiments using the acceleration sensor of smartphones [for description of the function and the use of the acceleration sensor, see [1] (Chap. 6)] within this column, in this case for analyzing the radial acceleration.

Radial acceleration is investigated in several experiments with smartphones: One experiment is performed with experimental apparatus in a physics laboratory; the other experiment is carried out with pupils at a children's playground. This second example provides a means of exploring radial acceleration using an everyday object—in this particular example a merry-go-round. In this contribution the same apps described in previous contributions about the use of acceleration sensors installed in smartphones [1–3] (Chaps. 6, 29 and 32) are used (SPARKvue [4] with an iPhone or an iPod touch, Accelogger [5] for an Android device). The values measured by the smartphone are also subsequently exported to a spreadsheet application (e.g., MS Excel) for analysis [1].

16.1 Radial Acceleration in the Physics Laboratory

In this example, a roof slat with a length of almost 2 m is fixed to an electric motor—as is often found in physics collections, e.g., for experiments with a "Kugelschwebe" (semi-circular channel) (Fig. 16.1). With the help of cable fixers, the smartphone is then fixed onto the wooden slat at a defined distance from the rotation center r so that the axis is pointing in the direction of radial acceleration [6]. Figure 16.2 shows a

P. Vogt (✉)
Institute of Teacher Training (ILF) Mainz, Mainz, Germany
e-mail: vogt@ilf.bildung-rp.de

J. Kuhn
Ludwig-Maximilians-Universität München (LMU Munich), Faculty of Physics, Chair of Physics Education, Munich, Germany
e-mail: jochen.kuhn@lmu.de

Fig. 16.1 Experiment setup
to investigate radial
acceleration

Fig. 16.2 Measurement
example for a distance from
the rotation center of
$r = 0.865$ m and a duration of
circulation of $t = 38.2$ s for
20 revolutions

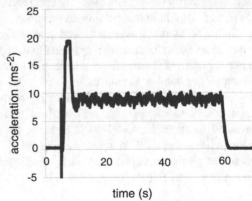

measurement example for a distance from the rotation center of 86.5 cm. When the
measurement is started, the motor is switched off; the measured radial acceleration is
close to zero (small deviations can arise because the smartphone is not positioned
perfectly horizontally). After approximately 5 s, the motor is switched on. From this
time onward, the iPhone moves with a constant track speed v.

If the acceleration values recorded at an interval of 12 s and 59 s are averaged, the
value is calculated to be 8.69 ms^{-2}. This result can be compared with a conventional
measurement, in which radial acceleration a is indirectly determined using the
formula

$$a = \frac{v^2}{r}. \tag{16.1}$$

For this, the time t is measured for a given number of revolutions n. For the
measurement example in Fig. 16.2, 20 revolutions occurred in 38.2 s. Taking into

consideration the formula for the circumference and Eq. (16.1), radial acceleration is calculated to be

$$a = \frac{n^2 4\pi^2 r}{t^2} = \frac{20^2 \cdot 4\pi^2 \cdot 0.865 \, \text{m}}{(38.5 \, \text{s})^2} \approx 9.36 \frac{\text{m}}{\text{s}^2}, \tag{16.2}$$

which matches well with the result of the smartphone measurement. Alongside individual measurements of radial acceleration, the setup makes it possible to verify Eq. (16.1) in an experiment. Namely, by recording a series of measurements, it is possible to confirm the proportionalities $a \approx v^2$ (for $r = $ constant) and $a \approx 1/r$ (for $v = $ constant). However, when selecting the velocity, it is advisable to limit the measurement range of the acceleration sensors installed in the smartphone to $\pm 2g$.

16.2 Centripetal Acceleration of a Merry-Go-Round

In the second example, the radial acceleration of a merry-go-round, typically found at children's playgrounds, is examined (Fig. 16.3). In order to make the carousel rotate, it is necessary to step onto it and apply force tangentially to the circular disc in the middle of the merry-go-round. Before the merry-go-round starts rotating, the smartphone is fixed at a given distance r from the rotation center either on the outside railing or the seating area of the merry-go-round so that an axis is pointing in the direction of radial acceleration. Similar to the previous example, this is performed with the help of cable fixers or adhesive tape, for example. Figure 16.4 shows a

Fig. 16.3 Experimental setup to examine centripetal acceleration of a merry-go-round

Fig. 16.4 Measurement
example from the merry-go-
round

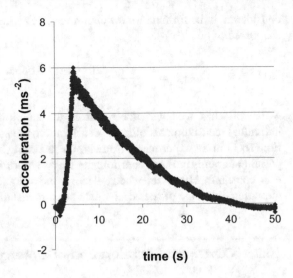

measurement example for a distance from the rotation center of 79 cm. After the
measurement has been started on the app, the merry-go-round is accelerated from a
complete standstill to a maximum value. In this example, the process lasts for
approximately 8 s. After this, the acceleration process is stopped and the merry-
go-round slows as a result of friction until it reaches a complete standstill (Fig. 16.4).

In order to obtain radial acceleration, several acceleration values recorded at a
short interval are averaged at the end of the acceleration process (e.g., at 9 s and 14 s;
Fig. 16.5). In this case, it results in a value of approximately 3.52 ms^{-2}. This result
can be compared to a conventional measurement in which radial acceleration a is
indirectly determined by applying Eq. (16.1). Then Eq. (16.2) is applied, resulting in
$a = 3.73$ ms^{-2} for $T = 2.89$ s and $r = 0.79$ m, so that the values from the experiments
can also be considered acceptable with a relative error of approximately 6%.

Studying this phenomenon using smartphones' acceleration sensor should also be
integrated in a more sophisticated instructional setting, connected with other phenomena [7–9] addressing the students' misconceptions which could be related to this
concept [10]. Anyway, current research shows that using this method to study free
fall and oscillation phenomena could at least increase curiosity and motivation of the
students when they learn with smartphones or tablets as experimental tools [11].

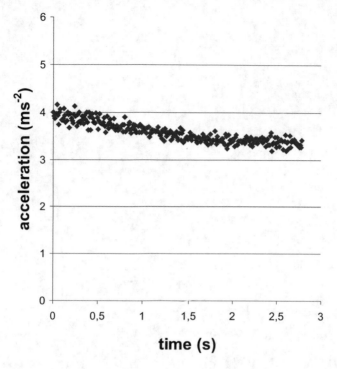

Fig. 16.5 Analysis using an iPhone

References

1. Vogt, P., Kuhn, J.: Analyzing free fall with a smartphone acceleration sensor. Phys. Teach. **50**, 182–183 (2012)
2. Vogt, P., Kuhn, J.: Analyzing simple pendulum phenomena with a smartphone acceleration sensor. Phys. Teach. **50**, 439–440 (2012)
3. Kuhn, J., Vogt, P.: Analyzing spring pendulum phenomena with a smartphone acceleration sensor. Phys. Teach. **50**, 504–505 (2012)
4. https://ogy.de/sparkvue
5. https://ogy.de/accelogger
6. In order to obtain a very precise specification of the distance to the rotation center, information on the location of the acceleration sensors within the smartphone must be obtained from the manufacturer. In the iPod touch (4G) the sensors are located underneath the home button
7. Kuhn, J.: Relevant information about using a mobile phone acceleration sensor in physics experiments. Am. J. Phys. **82**(2), 94 (2014)
8. Kuhn, J., Vogt, P., Müller, A.: Analyzing elevator oscillation with the smartphone acceleration sensors. Phys. Teach. **52**(1), 55–56 (2014)
9. Vogt, P., Kuhn, J.: Analyzing collision processes with the smartphone acceleration sensor. Phys. Teach. **52**(2), 118–119 (2014)
10. Hall, J.: iBlack Box? Phys. Teach. **50**(5), 260 (2012)
11. Hochberg, K., Kuhn, J., Müller, A.: Using smartphones as experimental tools – Effects on interest, curiosity and learning in physics education. J. Sci. Educ. Technol. **27**(5), 385–403 (2018)

Detect Earth's Rotation Using Your Smartphone

17

Julien Vandermarlière

If Galileo had had a smartphone... could he have proved that Earth rotates about its own axis? Perhaps! For that he could have used the accelerometers, which they all contain. Their reliability for carrying out scientific experiments has been tested numerous times [1–7]. Thanks to them we can measure the value of the acceleration due to gravity. It turns out that this value changes according to our latitude, mainly due to the centrifugal effect linked to this rotation. In this chapter we propose an easy method that can be used to detect this effect. It will suffice to measure the value of the acceleration due to gravity, which will be measured at different latitudes with a smartphone during a trip.

17.1   Theoretical Background

In 1672, Jean Richer [8], a French astronomer, was sent to Cayenne (French Guyana) to study the parallax of the planet Mars. During this trip he noticed that a pendulum beats slower near the equator than in Paris. This observation was quickly interpreted by Huygens as being an effect of the rotation of Earth! The closer we get to the equator, the more distant we are away from its axis of rotation. This increases the centrifugal acceleration to which we are subjected. As this acceleration is opposite the acceleration due to gravity, the closer we get to the equator, the lower is the apparent gravity! In addition, this centrifugal acceleration causes an equatorial bulge. This moves us a little bit further from the axis of rotation, thus increasing the centrifugal effect and lowering the value of the apparent gravity even more. Huygens and Newton quickly attempted to mathematically model this variation of the gravity as a function of latitude, and expeditions were sent worldwide to measure it. Nowadays, the World Geodetic System Ellipsoidal Gravity Formula (Eq. 17.1) [9]

J. Vandermarlière (✉)
Lycée Jean Lurçat, Perpignan, France
e-mail: julien.vandermarliere@ac-montpellier.fr

accounts for this phenomenon with good precision (where ϕ is the value of latitude at the considered place). In this chapter it will be used, thanks to an online calculator [10], to find out the theoretical values that will be compared to the experimental ones.

$$g_{(\text{latitude})} = 9.7803267714 \cdot \left(\frac{1 + 0.00193185138639 \cdot \sin^2(\phi)}{\sqrt{1 - 0.0066943799013 \cdot \sin^2(\phi)}} \right) \qquad (17.1)$$

17.2 Description of the Experiment

If one wants to detect these tiny variations of gravity with a smartphone, one should make a trip with a wide variation in latitude. This is what I was lucky to do on a flight between Cancun, Mexico (latitude 21.2°), and Chicago, USA (latitude 41.9°) (Fig. 17.1). These two locations have almost the same longitude and a very low altitude. In order to measure g, phyphox [11] was used. This application offers the possibility to start a recording with a "start delay" and with an automatic stop after a selected duration (Fig. 17.2). This does not disturb the measurements when the

Fig. 17.1 From Cancun to Chicago, the perfect trip to detect the rotation of Earth! (Credit Google Earth)

Fig. 17.2 Phyphox allows
timed run experiments

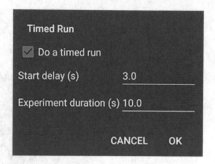

screen is touched. The method chosen was to leave the smartphone in my pants
pocket for at least 1 h (to ensure an almost constant temperature between
measurements), to put the smartphone horizontally on a soft seat (in order to avoid
vibrations), and then launch an acquisition of 10 s. The sampling frequency depends
mostly on the sensor embedded in the smartphone. Phyphox allows you to find out
its name and its characteristics. In my case, with a Huawei Mate 20, the sensor is an
InvenSense ICM-20690. The sampling frequency was 500 Hz. That makes 5000
measurements for 10 s. This is enough to make a significant average. By proceeding
this way, it was possible to obtain reproducible measurements of the average at more
or less 0.003 m/s^2. The standard deviation of the measurements (Fig. 17.3) was about
0.008 m/s^2 and the standard error of the average values 0.0001 m/s^2. It is important
to carry out the experiment in a short period of time because from 1 week to another
the calibration of the acceleration sensor can suffer from a drift. Equation (17.1)
predicts the difference of g between Chicago ($g = 9.803$ m/s^2) and Cancun
($g = 9.787$ m/s^2) to be 0.016 m/s^2; we will be comparing our experimental results
to this prediction.

17.3 Results

On Feb. 22, 2020, at Cancun airport, a value of $g = 9.741$ m/s^2 was found. Six hours
later, another measurement was carried out at the Chicago airport, resulting in
$g = 9.760$ m/s^2. The subtraction between these two values, 0.019 m/s^2 is in pretty
good agreement with the theoretical values given by Eq. (17.1) (Table 17.1).

17.4 Conclusion

It therefore seems possible to detect one of the effects of Earth's rotation thanks to a
smartphone! It is simply presented as a tool allowing one to perform in a practical
way the same experiments as those traditionally done in class. Galileo would surely
have been one of the great contributors to these columns if he had had a smartphone!

Fig. 17.3 An example of the measurements

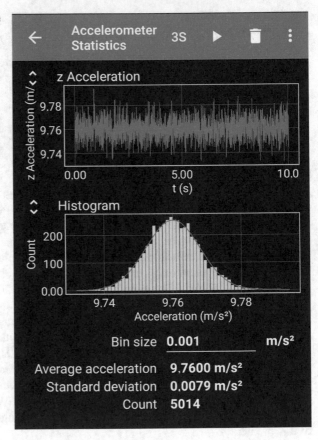

Table 17.1 Experimental and theoretical values of *g* measured with a single smartphone on February 22, 2020

Location	Theoretical values	Experimental values
Cancun (Mexico), 21.2°	9.787 m/s^2	9.741 ± 0.003 m/s^2
Chicago (USA), 41.9°	9.803 m/s^2	9.760 ± 0.003 m/s^2

References

1. Vogt, P., Kuhn, J.: Analyzing radial acceleration with a smartphone acceleration sensor. Phys. Teach. **51**, 182 (2013)
2. Vogt, P., Kuhn, J.: Analyzing free fall with a smartphone acceleration sensor. Phys. Teach. **50**, 182 (2012)
3. Kuhn, J., Vogt, P., Müller, A.: Analyzing elevator oscillation with the smartphone acceleration sensors. Phys. Teach. **52**, 55 (2014)
4. Hochberg, K., Becker, S., Louis, M., Klein, P., Kuhn, J.: Using smartphones as experimental tools – a follow-up: cognitive effects by video analysis and reduction of cognitive load by multiple representations. J. Sci. Educ. Technol. **29**(2), 303–317 (2020)

5. Hochberg, K., Kuhn, J., Müller, A.: Using smartphones as experimental tools – effects on interest, curiosity and learning in physics education. J. Sci. Educ. Technol. **27**(5), 385–403 (2018)
6. Becker, S., Gößling, A., Klein, P., Kuhn, J.: Using mobile devices to enhance inquiry-based learning processes. Learn. Instr. **69**, 101350 (2020)
7. Becker, S., Gößling, A., Klein, P., Kuhn, J.: Investigating dynamic visualizations of multiple representations using mobile video analysis in physics lessons: effects on emotion, cognitive load and conceptual understanding. Zeitschrift für Didaktik der Naturwissenschaften. **26**, 123–142 (2020)
8. National Imagery and Mapping Agency: NIMA TR 8350.2: Department of Defense World Geodetic System 1984, 3rd edn, (2000)
9. Richer, J.: The Galileo project. http://galileo.rice.edu/Catalog/NewFiles/richer.html
10. International Gravity Formula: https://www.vcalc.com/wiki/vCalc/International+Gravity+Formula
11. https://phyphox.org/

Angular Velocity and Centripetal Acceleration Relationship

18

Martín Monteiro, Cecilia Cabeza, Arturo C. Marti, Patrik Vogt, and Jochen Kuhn

During the last few years, the growing boom of smartphones has given rise to a considerable number of applications exploiting the functionality of the sensors incorporated in these devices. A sector that has unexpectedly taken advantage of the power of these tools is physics teaching, as reflected in several recent papers [1–10]. In effect, the use of *smartphones* has been proposed in several physics experiments spanning mechanics, electromagnetism, optics, oscillations, and waves, among other subjects. Although mechanical experiments have received considerable attention, most of them are based on the use of the accelerometer (e.g. Chaps. 6, 16, 29, and 32) [1–8]. An aspect that has received less attention is the use of *rotation sensors* or *gyroscopes* (e.g. Chaps. 22 and 39) [9, 10]. An additional advance in the use of these devices is given by the possibility of obtaining data using the accelerometer and the gyroscope simultaneously. The aim of this chapter is to consider the relation between the centripetal acceleration and the angular velocity. Instead of using a formal laboratory setup, in this experiment a smartphone is attached to the floor of a merry-go-round, found in many playgrounds. Several experiments were performed with the roundabout rotating in both directions

M. Monteiro (✉)
Universidad ORT Uruguay, Montevideo, Uruguay
e-mail: monteiro@ort.edu.uy

C. Cabeza · A. C. Marti
Universidad de la República, Montevideo, Uruguay
e-mail: cecilia@fisica.edu.uy; marti@fisica.edu.uy

P. Vogt
Institute of Teacher Training (ILF) Mainz, Mainz, Germany
e-mail: vogt@ilf.bildung-rp.de

J. Kuhn
Ludwig-Maximilians-Universität München (LMU Munich), Faculty of Physics, Chair of Physics Education, Munich, Germany
e-mail: jochen.kuhn@lmu.de

and with the smartphone at different distances from the center. The coherence of the measurements is shown.

18.1 Experimental Setup

The experimental setup, shown in Fig. 18.1, consists of a smartphone placed in a box made with polyurethane foam and fixed to the floor of the merry-go-round using two strong neodymium magnets. An LG Optimus P990 2X (Sensors: three-axis acceler-ometer KXTF9 Kionix, accuracy 0.001 ms^{-2}, three-axis gyroscope MPU3050 Invensense, accuracy 0.0001 rad/s) similar to the one used in [10]) was used. It was oriented with the display pointing upward and the short end parallel to the radial direction, as shown in the figure. Measurements that are relevant in this experiment are those reported by the rotation sensor according to the z-axis and the radial acceleration corresponding to the x-axis.

18.2 Rotatory Motion

The merry-go-round was propelled in a counterclockwise direction and allowed to come to a stop by the effect of friction, and then propelled again but in the clockwise direction. The experiment was repeated for different distances of the smartphone to the rotation center: 40, 60, 80, 100, and 120 cm. The angular velocity is measured

Fig. 18.1 Smartphone mounted on a merry-go-round

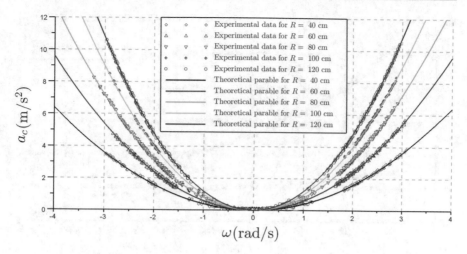

Fig. 18.2 Centripetal acceleration as a function of the angular velocity for different distances indicated in the legend box

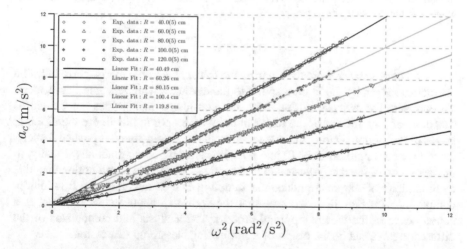

Fig. 18.3 Centripetal acceleration as a function of the angular velocity squared for different distances

with the z-component of the gyroscope, while the centripetal acceleration is measured with the x-axis of the accelerometer.

The measurements obtained are summarized in Figs. 18.2 and 18.3, where the centripetal acceleration a_c is plotted as a function of the angular velocity ω and the angular velocity squared ω^2, respectively. The linear and parabolic fits included in these figures reveal that both magnitudes are related by the well-known relationship

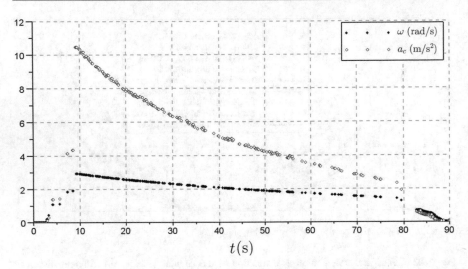

Fig. 18.4 Angular velocity and centripetal acceleration as a function of time, corresponding to one of the realizations shown in the previous figures. The distance is $R = 120$ cm and the merry-go-round is spinning counterclockwise

$$a_c = \omega^2 R, \tag{18.1}$$

where R is the distance from the smartphone to the axis of the merry-go-round. The coefficients given by the fit correspond to the distances with very good agreement. To complete the analysis in Fig. 18.4, the considered magnitudes are plotted as functions of time for one of the realizations. We observe in this figure the different stages of the motion. During the first seconds, the merry-go-round is pushed fiercely. Next, between approximately 10 and 80 s, it is slowing down gradually. Finally, in the last seconds, the merry-go-round is abruptly stopped. It is worth noting that, due to limitations of the smartphone, the sampling rate is not uniform. In addition, comparing with Fig. 18.3, we note that the wide gap about $\omega^2 \approx 2$ rad^2/s^2 is a consequence of the violent stopping process. The analysis and comparison of the different figures can be the origin of a stimulating classroom discussion.

18.3   Conclusion

A basic kinematic relationship between angular velocity and centripetal acceleration was verified using smartphone sensors. The coherence of the measurements taken with the different sensors was shown. This experiment illustrates the simplicity of using a smartphone in physics experiments. It is worth mentioning that the experiment proposed here is not easy to implement in a traditional laboratory. Indeed, angular velocity measurements require rotation sensors that are not easily coupled to rotating devices such as a merry-go-round. In addition, traditional sensors available in most laboratories are not only considerably more expensive than smartphones, but also need wired connections. A similar experimental setup, without using smartphones, is far more complex than that proposed in this chapter.

The experiment could also be conducted in a classroom, e.g., with a rotating disk, if no adequate merry-go-round is available.

References

1. Vogt, P., Kuhn, J., Müller, S.: Experiments using cell phones in physics classroom education: the computer-aided g determination. Phys. Teach. **49**, 383 (2011)
2. Vogt, P., Kuhn, J.: Analyzing free fall with a smartphone acceleration sensor. Phys. Teach. **50**, 182 (2012)
3. Vogt, P., Kuhn, J.: Analyzing simple pendulum phenomena with a smartphone acceleration sensor. Phys. Teach. **50**, 439–440 (2012)
4. Kuhn, J., Vogt, P.: Analyzing spring pendulum phenomena with a smartphone acceleration sensor. Phys. Teach. **50**, 504 (2012)
5. Streepey, J.W.: Using iPads to illustrate the impulse momentum relationship. Phys. Teach. **51**, 54 (2013)
6. Vogt, P., Kuhn, J.: Analyzing radial acceleration with a smartphone acceleration sensor. Phys. Teach. **51**, 182 (2013)
7. Kuhn, J., Vogt, P.: Smartphones as experimental tools: different methods to determine the gravitational acceleration in classroom physics by using everyday devices. Eur. J. Phys. Educ. **4**, 16 (2013)
8. Chevrier, J., Madani, L., Ledenmat, S., Bsiesy, A.: Teaching classical mechanics using smartphones. Phys. Teach. **51**, 376 (2013)
9. Shakur, A., Sinatra, T.: Angular momentum. Phys. Teach. **51**, 564 (2013)
10. Monteiro, M., Cabeza, C., Martí, A.C.: Rotational energy in a physical pendulum. Phys. Teach. **52**, 561 (2014)

Determination of the Radius of Curves and Roundabouts with a Smartphone

19

Christoph Fahsl and Patrik Vogt

Based on earlier work [1], this chapter describes two further experiments that can be carried out on the road. It will be explained how to determine the radius of curves and roundabouts of public streets using only a smartphone. The first experiment shows how to determine the radius of a curve by driving a car around the curve while sampling the acceleration data of the car. The second experiment shows how to calculate the radius of a roundabout by using the built-in gyroscope sensor in combination with the acceleration sensor of the smartphone. The same procedure was used by Monteiro et al. (Chap. 18) [2] to examine a merry-go-round.

19.1 Determination of a Curve Radius Using the Acceleration Sensors

For this experiment, the smartphone has to be mounted in the car, in order that the two acceleration sensors are correctly aligned—one in the direction of motion and one orthogonal to the direction of motion. To execute the experiment, start the sampling of the acceleration data while the car is in a standstill position. When the car is at a standstill, accelerate to a certain velocity that permits one to move around the bend at a constant speed and radius. The car, as well as the measurement, can be stopped once the curve has been passed. In the following graph you can see the acceleration data as a function of time (Fig. 19.1).

By integrating the acceleration data in the direction of motion over time, you obtain the velocity graph of the car as a function of time (Fig. 19.2).

C. Fahsl (✉)
Bertolt-Brecht-Schule (High-School), Nürnberg, Germany

P. Vogt
Institute of Teacher Training (ILF) Mainz, Mainz, Germany
e-mail: vogt@ilf.bildung-rp.de

© The Author(s), under exclusive license to Springer Nature Switzerland AG 2022
J. Kuhn, P. Vogt (eds.), *Smartphones as Mobile Minilabs in Physics*,
https://doi.org/10.1007/978-3-030-94044-7_19

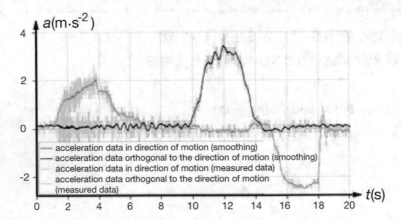

Fig. 19.1 The acceleration data as functions of time

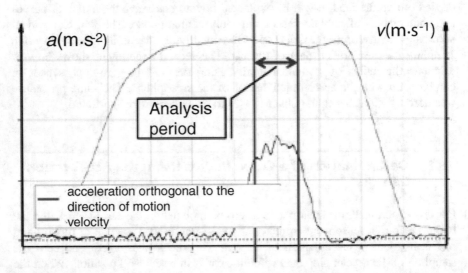

Fig. 19.2 The velocity of the car and the acceleration data orthogonal to the car as functions of time

The relation between velocity and orthogonal acceleration (centripetal acceleration) of a constant circular motion allows us to determine the radius of the curve:

$$a_c = \frac{v^2}{r}$$

(a_c = orthogonal acceleration, v = velocity of the car, r = radius of curve). The following section shows the result of three consecutive measurements including the standard errors of the average means:

Fig. 19.3 Google Maps overview of the curve used in this experiment

$$r_1 = (14.1 \pm 0.1)\text{m}, \quad r_2 = (14.6 \pm 0.1)\text{m}, \quad r_3 = (13.61 \pm 0.12)\,\text{m}.$$

The analysis of the curve via Google Maps yields to a radius of 13.7 m and hence agrees with the values determined by the experiment (Fig. 19.3).

19.2 Determination of the Radius of a Roundabout Using the Acceleration Sensor in Combination with the Gyroscope Sensor

For this experiment, the smartphone has to be aligned in the direction of movement. Afterwards, you sample the data of the gyroscope and acceleration sensor simultaneously while driving around a roundabout with constant speed and radius (Fig. 19.4).

The relation between centripetal force and angular velocity allows the calculation of the radius of the roundabout (Fig. 19.4):

$$r = \frac{a}{\omega^2}$$

(r = radius, a = centripetal acceleration, ω = angular velocity). Two consecutive measurements lead to the following results

$$r_1 = (11.42 \pm 0.01)\,\text{m}, \quad r_2 = (11.789 \pm 0.019)\,\text{m}.$$

The analysis of the curve via Google Maps yields to a radius of 11.2 m (see Fig. 19.5). Again, the values of the experiment are consistent with the theoretical values determined by the analysis of the roundabout via Google Maps.

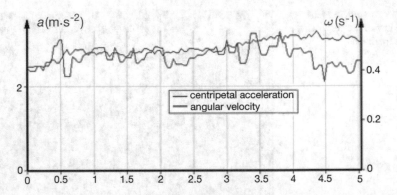

Fig. 19.4 The centripetal acceleration and angular velocity as functions of time

Fig. 19.5 Google Maps
overview of the roundabout
used in this experiment

Finally it can be concluded that with the two presented methods, curve radii can be determined with sufficient accuracy for educational purposes.

References

1. Fahsl, C., Vogt, P.: Determination of the drag resistance coefficients of different vehicles. Phys. Teach. **56**, 324–325 (2018)
2. Monteiro, M., Cabeza, C., Marti, A.C., Vogt, P., Kuhn, J.: Angular velocity and centripetal acceleration relationship. Phys. Teach. **52**, 312–313 (2014)

Understanding Coffee Spills Using a Smartphone

20

Fernando Tornaría, Martín Monteiro, and Arturo C. Marti

The SpillNot® is an ingenious and effective device that aims to solve the everyday problem of transporting an open cup of hot beverage like tea or coffee without spilling. It not only avoids spills under the normal conditions in which a drink is usually carried, but also remains effective in extreme conditions such as giving full turns in a vertical or a horizontal plane. To help explain the operation of this device, instead of a cup, a smartphone was placed on the base of a SpillNot®. The acceleration components, parallel and perpendicular to the base, were obtained using the built-in accelerometer. The analysis of these measures sheds light on the physical mechanisms of the SpillNot®.

20.1   The Physics of the SpillNot®

Everyone has experienced the problem of carrying a cup containing a hot beverage like coffee or tea. Common knowledge suggests that one must be especially careful to avoid spills. One possible solution is to use a SpillNot [1], a simple device aimed at solving this daily problem. It consists of an anti-slip pad placed on a base that is attached by a handle, which in turn is held by a ribbon. When someone carries a SpillNot® held by the ribbon with a cup placed on its base, it is possible to oscillate the device with large amplitudes and even to give full turns in a vertical or in a horizontal plane, as shown in Fig. 20.1.

F. Tornaría (✉)
CES-ANEP, Montevideo, Uruguay

M. Monteiro
Universidad ORT Uruguay, Montevideo, Uruguay
e-mail: monteiro@ort.edu.uy

A. C. Marti
Universidad de la República, Montevideo, Uruguay
e-mail: marti@fisica.edu.uy

© The Author(s), under exclusive license to Springer Nature Switzerland AG 2022
J. Kuhn, P. Vogt (eds.), *Smartphones as Mobile Minilabs in Physics*,
https://doi.org/10.1007/978-3-030-94044-7_20

Fig. 20.1 Snapshot showing
a cup on a SpillNot® giving
full turns in a vertical plane;
the surface of the liquid can be
appreciated

Let us consider a simple model of the SpillNot® as a pendulum. On one hand,
defining θ as the angle with the vertical from the lowest position, the vertical and
horizontal acceleration components are given by

$$a_y = \frac{T}{m}\cos\theta - g$$

and

$$a_x = \frac{T}{m}\sin\theta$$

thus the angle of the pendulum with the vertical verifies

$$\tan\theta = \frac{a_x}{g + a_y}.$$

On the other hand, in general, a fluid cannot sustain a force that is tangential to its
surface. As a consequence, the free surface in a system subject to a constant
acceleration forms an angle with the horizontal, whose tangent is given by

$$\frac{a_x}{g + a_y},$$

where a_x and a_y are the horizontal and vertical acceleration components, as indicated
in Fig. 20.2. Putting these facts together, we see that, if we want no spilling, the angle
of the SpillNot® with respect to the *vertical* needs to be the same as the angle of the
liquid with respect to the *horizontal*. So, if the acceleration is approximately
perpendicular to the base, the liquid will remain roughly parallel to the base. This
analysis is approximate in several factors; transient effects, the motion of the hand, or
the finite size of the device, among others, are neglected. However, the key point,

Fig. 20.2 The slope of a free surface in a container moving with uniform acceleration a is perpendicular to direction of the sum of the gravitational and the fictitious force F^*

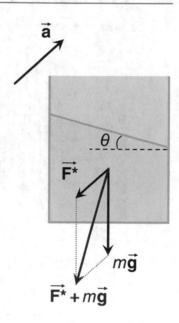

i.e., that the radial acceleration should be considerably larger than the tangential acceleration in order to limit spilling, is revealed.

20.2 Experimental Results

To verify this conclusion, instead of a cup of hot beverage, a smartphone was placed on the base of a SpillNot® and the system set in motion oscillating with a large amplitude in a vertical plane. An Android smartphone (Samsung GT-I9100) furnished with a three-axis accelerometer (ST-Microelectronics K3DH, 0.005 m/s² resolution and 16 Hz sampling rate) was used. The Androsensor application [2] was used to record sensor readings. The radial acceleration is given by the sensor along the z-axis while the vectorial sum of the values corresponding to the x- and y-axis is the tangential component.

The temporal evolution of the acceleration is shown in Fig. 20.3. It can be appreciated that the magnitude of the radial component is much greater than the tangential component. It must be highlighted that acceleration sensors measure, in fact, an apparent acceleration (Chap. 6) [3, 4], resulting from the vectorial sum of the real acceleration and the acceleration associated with a gravitational field in the opposite direction to that of the real gravitational field. As a consequence, the radial acceleration is centered about 10 m/s², instead of about a null value. In addition, the tangential component also oscillates around a non-zero value. This is due to the fact that the SpillNot® is asymmetric and, when hanging at rest held from the ribbon, the base is not horizontal but inclined at an angle of approximately 8°.

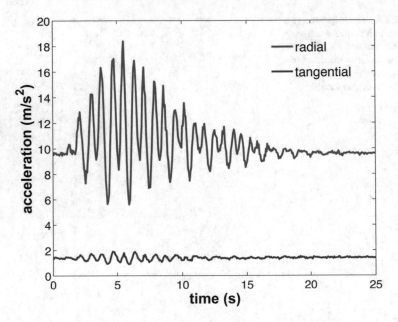

Fig. 20.3 Acceleration components as functions of time. During the interval between approximately 3 s and 15 s, the system is performing large oscillations in a vertical plane

20.3   Final Remarks

In this contribution a simple experiment using a smartphone helps to analyze a daily problem and discuss the underlying mechanisms of a device that, at first sight, appears to be magic. It is worth noting that, far from trivial, sloshing dynamics is a problem that goes beyond the carrying of hot beverages and has received considerable attention in physics and engineering [5]. For instance, one important application is the control of large liquid-filled structures such as rocket fuel tanks.

References

1. http://thespillnot.com
2. http://www.fivasim.com/androsensor.html
3. Vogt, P., Kuhn, J.: Analyzing free fall with a smartphone acceleration sensor. Phys. Teach. **50**, 182 (2012)
4. Monteiro, M., Cabeza, C., Marti, A.C.: Acceleration measurements using smartphone sensors: dealing with the equivalence principle. Rev. Bras. Ens. Fis. **37**(1), 1303 (2014)
5. Ibrahim, R.A.: Liquid Sloshing Dynamics: Theory and Applications. Cambridge University Press (2005)

Tilting Motion and the Moment of Inertia of the Smartphone

A. Kaps and F. Stallmach

Smartphones and their internal sensors offer new options for an experimental access to teach physics at secondary schools and universities. Especially in the field of mechanics, a number of smartphone-based experiments are known illustrating, e.g., linear and pendulum motions [1–4] as well as rotational motions [5–8] using the internal MEMS accelerometer and gyroscope, respectively.

In this chapter we propose to measure the angular velocity of the smartphone during a controlled tilting motion. The aim is to determine the moment of inertia of the smartphone and compare it to reference data provided by the theory of rigid body motions. The experiment itself takes just a few seconds. It requires a soft mat and a smartphone with an application recording the angular velocity during the tilting motion. For this purpose, we chosen the app *phyphox* [9] (RWTH Aachen, Germany) available for Android and iOS smartphones.

21.1 Theoretical Background

When the smartphone overturns from a position, where it freely stands with its center of mass S in the height d just above its longest edge b, onto a position, where it arrives with its back face on the soft mat (Fig. 21.1), its potential energy is transformed in kinetic energy of rotation [10]

$$\frac{I_b}{2} \cdot \omega_{y,\,max}^2 = m \cdot g \cdot \left(d - \frac{c}{2}\right). \tag{21.1}$$

A. Kaps (✉) · F. Stallmach
Department Didactics of Physics, Faculty for Physics and Earth Sciences, University Leipzig, Leipzig, Germany
e-mail: andreas.kaps@uni-leipzig.de

J. Kuhn, P. Vogt (eds.), *Smartphones as Mobile Minilabs in Physics*,
https://doi.org/10.1007/978-3-030-94044-7_21

Fig. 21.1 Sketch of the experiment. The smartphone tilts and impinges on a soft mat (drawn in orange) to inhibit sliding and to damp mechanical shocks

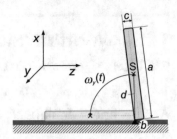

Here, m and I_b denote the mass and the moment of inertia of the smartphone, respectively, g is the gravitational acceleration, and $\omega_{y,max}$ represents the angular velocity of the circular motion in the moment of time the smartphone meets the mat.

Because the axis of rotation is the edge b (Fig. 21.1), the length d that follows from the dimensions of the smartphone via the Pythagorean theorem

$$d = \frac{1}{2}\sqrt{a^2 + c^2} \tag{21.2}$$

is also required for calculating the moment of inertia I_b using the parallel axis theorem [10],

$$I_b = I_y + m \cdot d^2. \tag{21.3}$$

I_y is the principal moment of inertia of the smartphone for an axis pointing in the y-direction. Combining Eqs. (21.1) to (21.3) we find

$$I_y = \frac{m \cdot g \cdot (2 \cdot d - c)}{\omega_{y,\,max}^2} - m \cdot d^2. \tag{21.4}$$

A suitable reference value for comparison with the experimental result obtained via Eq. (21.4) may be calculated by assuming the smartphone as a homogeneous cuboid [6]. The corresponding principal moment of inertia is

$$I_y = \frac{1}{12} \cdot m \cdot (a^2 + c^2), \tag{21.5}$$

where a and c are the lengths of the two edges of the smartphone orientated perpendicular to the y-axis.

21.2   The Experiment

The experimental setup is depicted in Fig. 21.1. The smartphone initially stands on a non-slippery soft mat. The app *phyphox* records and displays the data of the angular velocity $\omega_y(t)$ from the MEMS gyroscope. The smartphone is carefully moved until its center of mass S is directly above its long edge b. In this moment it starts to tilt

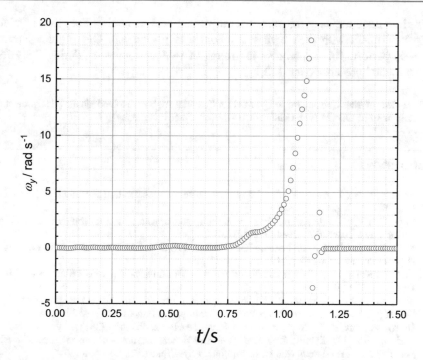

Fig. 21.2 Angular velocity $\omega_y(t)$ of the smartphone during the tilting experiment. The maximum value is $\omega_{y,\mathrm{max}} = (18.5 \pm 0.4)$ rad $\cdot$ s^{-1}

Table 21.1 Data of the smartphone as given by the manufacturer [11] and the experimental [Eq. (21.4)] and reference [Eq. (21.5)] values of the moment of inertia I_y

length a (x-direction)	(7.81 ± 0.01) cm
height b (y-direction)	(15.84 ± 0.01) cm
width c (z-direction)	(0.75 ± 0.01) cm
mass m	(239 ± 1) g
experimental value I_y	$(1.18 \pm 0.35) \cdot 10^{-4}$ kg m^2
reference value I_y	$(1.23 \pm 0.05) \cdot 10^{-4}$ kg m^2

freely under the influence of the gravitational force. After it comes to rest on the mat, the data recording is stopped and the $\omega_y(t)$ data may be analyzed or exported to another device.

Figure 21.2 displays the angular velocity $\omega_y(t)$. The onset of the free tilting at $t \approx 0.75$ s, the maximum angular velocity ($t_{\mathrm{max}} \approx 1.13$ s), as well as the impingement of the smartphone on the mat ($t \geq 1.15$ s) are easily identified. We just read the maximum angular velocity from the data table or the graph (Fig. 21.2), estimate its measurement uncertainty, and calculate the experimental and the reference values of the moment of inertia of the smartphone with Eqs. (21.4) and (21.5), respectively (see Table 21.1). Both values are found to agree within the given measurement uncertainties.

In summary, the proposed experiment yields reasonable quantitative results for the principal moment of inertia of the smartphone. For students, it replicates and joins important topics of the rotational movement of rigid bodies. They may conduct this experiment with their own smartphones as pocket lab. Our physics teacher trainees perform this experiment in groups of two as homework during their first experimental physics course [12].

Acknowledgments The authors are grateful for financial support received via the STIL project of the University of Leipzig (grant number 01PL16088, BMBF Germany).

References

1. Vogt, P., Kuhn, J.: Analyzing free fall with a smartphone acceleration sensor. Phys. Teach. **50**, 182 (2012)
2. Vogt, P., Kuhn, J.: Analyzing simple pendulum phenomena with a smartphone acceleration sensor. Phys. Teach. **50**, 439–440 (2012)
3. Fahsl, C., Vogt, P.: Determination of the radius of curves and roundabouts with a smartphone. Phys. Teach. **57**, 566–567 (2019)
4. Vogt, P., Kuhn, J.: Analyzing radial acceleration with a smartphone acceleration sensor. Phys. Teach. **51**, 182 (2013)
5. Kaps, A., Splith, T., Stallmach, F.: Shear modulus determination using the smartphone in a torsion pendulum. Phys. Teach. **59**, 268 (2021). https://doi.org/10.1119/10.0004154
6. Shakur, A., Sinatra, T.: Angular moment. Phys. Teach. **51**, 564–565 (2013)
7. Patrinopoulos, M., Kefalis, C.: Angular velocity direct measurement and moment of inertia calculation of a rigid body using a smartphone. Phys. Teach. **53**, 564–565 (2015)
8. Monteiro, M., Cabeza, C., Marti, A.: Rotational energy in a physical pendulum. Phys. Teach. **52**, 312–313 (2014)
9. The phyphox homepage by the RTWH Aachen. https://phyphox.org/de/home-de/
10. Halliday, D., Resnick, R., Walker, J.: Fundamentals of Physics, 9th edn. Wiley (2010)
11. We used the mass and the dimensions given from the technical data sheet by Apple Inc. https://www.apple.com/de/iphone/compare/
12. Kaps, A., Splith, T., Stallmach, F.: Implementation of smartphone-based experimental exercises for physics courses at universities. Phys. Educ. **56**(3), 0035004. https://doi.org/10.1088/1361-6552/abdee2

Angular Momentum

22

Asif Shakur and Taylor Sinatra

The gyroscope in a smartphone was employed in a physics laboratory setting to verify the conservation of angular momentum and the nonconservation of rotational kinetic energy. As is well-known, smartphones are ubiquitous on college campuses. These devices have a panoply of built-in sensors. This creates a unique opportunity for a new paradigm in the physics laboratory (Chaps. 6, 29, and 32) [1–3]. Many traditional physics experiments can now be performed very conveniently in a pedagogically enlightening environment while simultaneously reducing the laboratory budget substantially by using student-owned smartphones.

22.1 Experimental Procedure

A "Lazy Susan" turntable was acquired very costeffectively [4] and used as a rotating platform. A smartphone was secured to the edge of the turntable by means of masking tape (Fig. 22.1). The "record" button on the app, xSensor by Crossbow Technology Inc. [5], was pressed and the turntable was manually spun subsequently and almost simultaneously. The time delay does not have a deleterious effect on the integrity of the measurements, as the following discussion will convince the reader. The gyroscope output from the smartphone was recorded by the app for 5 s.

This provided us the "control" data for the normal deceleration of the angular velocity ω_z (gyroscope output of the smartphone), of the manually spun turntable. The weight was a 2.5-lb (1.13-kg) disk. The next step was to spin the turntable manually and gently drop the weight onto the spinning turntable. The experiment was repeated for three different positions where the weight was dropped. The gyroscope app on the smartphone dutifully recorded the angular velocity every

A. Shakur (✉) · T. Sinatra
Salisbury University, Salisbury, MD, USA
e-mail: amshakur@salisbury.edu

Fig. 22.1 Smartphone secured on a "Lazy Susan" turntable

Table 22.1 Gyroscope data for disk drop at three different distances on the Lazy Susan

	ω_i	ω_f	$I_i\omega_i$	$I_f\omega_f$	$\frac{1}{2} I_i \omega_i^2$	$\frac{1}{2} I_f \omega_f^2$
$r = 0$ cm	11.9	10.1	0.199	0.193	1.19	0.98
$r = 5$ cm	11.5	8.8	0.192	0.193	1.1	0.85
$r = 10$ cm	11.1	6.18	0.186	0.188	1.03	0.58

0.25 s. These data were analyzed to establish the conservation of angular momentum and the nonconservation of rotational kinetic energy.

22.2 Experimental Data

The gyroscope data for the 1.13-kg weight drop for three different positions of the drop ($r = 0$ cm, $r = 5$ cm, and $r = 10$ cm) are recorded in Table 22.1. The control data and one set of data for the 10-cm disk drop are depicted in Fig. 22.2.

22.3 Experiment Meets Theory

We will analyze the tabulated data and perform a sample calculation. But first let us get a few preliminaries out of the way. The lazy Susan has a mass of 415 g with a radius of 20 cm. This includes a black outer ring. The moment of inertia of the lazy Susan is calculated to be $12.63 \cdot 10^{-3}$ kg·m^2. The moment of inertia of the smartphone with case was calculated to be $4.13 \cdot 10^{-3}$ kg·m^2. Thus, the total moment of inertia of the lazy Susan and smartphone combo is $16.73 \cdot 10^{-3}$ kg·m^2. The 1.13-kg disk weight has a radius of 6.5 cm. We used the parallel-axis theorem to calculate the moment of inertia of the disk weight to be $2.43 \cdot 10^{-3}$ kg·m^2 when dropped at the 0-cm mark (center of lazy Susan), $5.23 \cdot 10^{-3}$ kg·m^2 when dropped at the 5-cm mark, and at the 10-cm mark to be $13.73 \cdot 10^{-3}$ kg·m^2. In Fig. 22.2 for the 1.13-kg disk weight drop at the 10-cm mark, we note a sudden drop in the angular velocity from

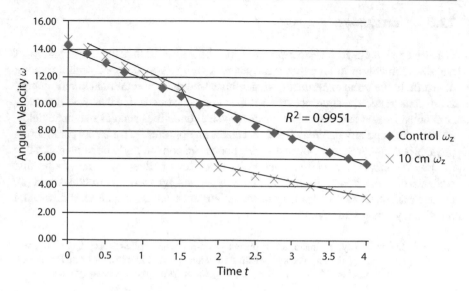

Fig. 22.2 Angular velocity vs time for the 1.13-kg weight drop

11.11 rad/s to 5.68 rad/s. Incorporating the control correction of 0.5 rad/s (which is how much the lazy Susan would have decelerated even without the weight drop), we reckon that the weight drop slowed the lazy Susan from 11.11 rad/s to 6.18 rad/s. The angular momentum before the weight drop is $I_i\omega_i = (16.73 \cdot 10^{-3})$ (11.11), so $I_i\omega_i = 0.186$. The angular momentum after the weight drop is $I_f\omega_f = (30.43 \cdot 10^{-3})$ (6.18), so $I_f\omega_f = 0.188$. So, the angular momentum is conserved to within 1%.

22.4   Is the Rotational Kinetic Energy Conserved?

We have established in the previous section that the angular momentum is conserved when we drop a weight onto a rotating platform. Let us consider the initial and final rotational kinetic energies for the same data from the gyroscope of the smartphone. In Fig. 22.2 for the 1.13-kg disk weight drop at the 10-cm mark, we note a sudden drop in the angular velocity from 11.11 rad/s to 5.68 rad/s. Incorporating the control correction of 0.5 rad/s (which is how much the lazy Susan would have decelerated even without the weight drop), we reckon that the weight drop slowed the lazy Susan from 11.11 rad/s to 6.18 rad/s. The rotational kinetic energy before the weight drop is $\frac{1}{2}I_i\,\omega_i^2 = \frac{1}{2}\,(16.73 \cdot 10^{-3})\,(11.11)^2$, so $\frac{1}{2}I_i\,\omega_i^2 = 1.03$ J. The rotational kinetic energy after the weight drop is $\frac{1}{2}\,I_f\,\omega_f^2 = \frac{1}{2}\,(30.43 \cdot 10^{-3})\,(6.18)^2$, so $\frac{1}{2}\,I_f\,\omega_f^2 = 0.581$ J. So, there is a 44% loss in the rotational kinetic energy.

22.5   Conclusion

We used the output of a smartphone gyroscope to establish the conservation of angular momentum in an experiment where a weight was dropped onto a rotating platform. In the same experiment, we demonstrated that the rotational kinetic energy is *not* conserved. The smartphone is a robust and versatile device that can accurately, conveniently, and reproducibly measure physical quantities such as magnetic fields, acceleration, and angular velocity. This creates an opportunity for a new paradigm in physics pedagogy. Student-owned smartphones can conveniently be implemented in the physics laboratory while reducing the laboratory budget. We have found that students take enormous pride in the data generated by their smartphones and are excited and motivated to learn from them. We even let them take their newfound physics toy home with them!

Acknowledgments   Taylor Sinatra, a very recent physics graduate of Salisbury University, was financially supported by the physics department. Both authors would like to express their indebtedness to Andrew Pica, chair of the physics department, for his support and encouragement.

References

1. Vogt, P., Kuhn, J.: Analyzing free fall with a smartphone acceleration sensor. Phys. Teach. **50**, 182–183 (2012)
2. Vogt, P., Kuhn, J.: Analyzing simple pendulum phenomena with a smartphone acceleration sensor. Phys. Teach. **50**, 439–440 (2012)
3. Vogt, P., Kuhn, J.: Analyzing radial acceleration with a smartphone acceleration sensor. Phys. Teach. **51**, 182–183 (2013)
4. Oxo Good Grips "Lazy Susan" turntable for $17 on Amazon
5. https://ogy.de/x-sensor

Angular Velocity Direct Measurement and Moment of Inertia Calculation of a Rigid Body Using a Smartphone

Matthaios Patrinopoulos and Chrysovalantis Kefalis

In this chapter we focus on smartphones as experimental tools; specifically we use the gyroscope sensor of a smartphone to study the turning motion of a rigid body. Taking into consideration recent work concerning that topic (Chaps. 18 and 39) [1, 2], we try to use the gyroscope sensor in studying the complex motion of a rolling cylinder on a slope.

Tablets and smartphones are very common now in our everyday life. These devices are equipped with a number of sensors, such as accelerometer, gyro sensor, thermo sensor, global positioning system sensor, light sensor, proximity, etc. Several works have proposed the use of smartphones in the conduction of laboratory experiments on mechanics (Chaps. 6, 10, 16, 22, 29, and 32) [3–11], electromagnetism (Chap. 58) [12, 13], optics (Chap. 66) [14], oscillations [15, 16], and waves (Chaps. 43 and 48) [17, 18].

23.1 Experimental Setup

In the experiment we used a rigid body (a barbell, consisting of two connected iron disks) rolling down an inclined plane in order to take direct measurements of its angular velocity during the motion. We attached the smartphone to the body near the axis of rotation. In this way, due to the small mass and dimensions of the mobile phone relative to these of the iron disks, we can take measurements without affecting the body's moment of inertia (Fig. 23.1) significantly.

According to our calculations, the moment of inertia of the used mobile phones (i.e., Samsung S3) proves to be two orders of magnitude smaller than that of the used rigid body. Specifically the moment of inertia of the mobile phone about the axis of

M. Patrinopoulos (✉) · C. Kefalis
Science, Technology and Environment Laboratory, Pedagogical Department of Primary Education, National and Kapodistrian University of Athens, Athens, Greece
e-mail: mpatrin@sch.gr; vkefalis@iit.demokritos.gr

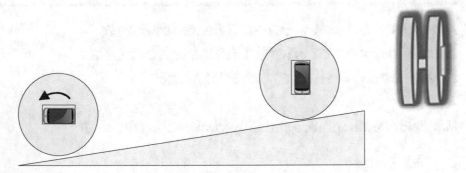

Fig. 23.1 Smartphone mounted on the center of a rigid body

rotation is $\approx 2.62 \cdot 10^{-4}$ kg·m^2, while the calculated moment of inertia of the rigid body (with the smartphone attached) was about $6.3 \cdot 10^{-2}$ kg·m^2 (our uncertainty in this estimate is unclear because the rigid body deviates from cylindrical shapes in unspecified ways, but even so, two orders of magnitude separate the two). The mass of the structure was 10.300 ± 0.005 kg and the corresponding radius 0.1100 ± 0.0005 m.

We used the application Physics Toolbox Gyroscope [19] to record the measurements. The smartphone we used was a Samsung S3, which is equipped with LSM330DLC gyroscope sensor, with limit ± 2000 deg/s. In this way (by using the smartphone as a laboratory instrument and its software), we can take measurements with a refresh rate up to 200 Hz. In order to carry out our measurements but also to conserve energy to be able to use the smartphone for a large number of other measurements, we chose to take measurements with the refresh rate of 20 Hz. These data can then be exported to a file and be analyzed by appropriate software. To verify the accuracy of the measurements from the smartphone, we also used video analysis of the motion. For a small-angle inclined plane ($\approx 2°$) we compared the data of the video analysis with the measurements we took from the smartphone and we could clearly see that there was no deviation of the angular velocity measured. This concurrence gave us confidence in the smartphone technique. At larger angles, such as the inclined plane of length 0.7000 ± 0.0005 m and height 0.0840 ± 0.0005 m considered below, video analysis did not allow the precision obtained below, which highlights an advantage of the smartphone method.

23.2   Measurements

While letting the body roll on a small-angle inclined plane, we record its angular velocity ω as function of time t. Our measurements are depicted on the diagram of Fig. 23.2. From this diagram we can clearly see that the angular velocity is proportional to the time as the rigid body accelerates, something that is confirmed by the theory.

From the diagram we proceed to calculate the angular acceleration:

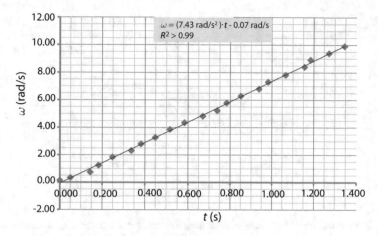

Fig. 23.2 Presentation of the measurements after exporting the data from the smartphone to MS Excel; from the slope of the diagram we can calculate the angular acceleration

$$a_{\mathrm{r}} = \frac{\omega_{\mathrm{bot}} - \omega_{\mathrm{up}}}{\Delta t}. \tag{23.1}$$

Based on the angular acceleration and given that the rigid body does not slide, we calculate the system's moment of inertia through the following relations:

$$m \cdot g \cdot \sin \theta - F_{\mathrm{s}} = m \cdot a_{\mathrm{cm}} \tag{23.2}$$

$$F_{\mathrm{s}} \cdot R = I \cdot a_{\mathrm{r}} \tag{23.3}$$

$$R \cdot a_{\mathrm{r}} = a_{\mathrm{cm}} \tag{23.4}$$

Thus the moment of inertia is given by the expression

$$I = \frac{R \cdot m \cdot g \cdot \sin \theta - m \cdot a_{\mathrm{r}} \cdot R^2}{a_{\mathrm{r}}}, \tag{23.5}$$

where:
m = the body's mass
a_{r} = the angular acceleration
a_{cm} = the acceleration of the body's center of mass
I = the moment of inertia
F_{s} = the friction
R = the cylinder's radius
θ = the inclined plane's angle

The angle of incline was about $7°$, and the angular acceleration as calculated from the trend line of the diagram is $a_{\mathrm{r}} = 7.43$ rad/s^2. The experimental value of the moment of inertia was found to be $(5.78 \pm 0.08) \cdot 10^{-2}$ kg·m^2, indicating that our theoretical value from above was an overestimate (a deviation of $\approx 8\%$).

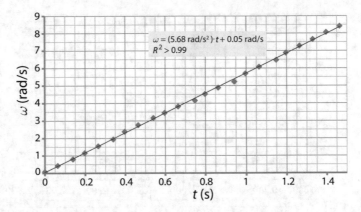

Fig. 23.3 Presentation of the measurements after exporting the data from the smartphone to MS Excel; from the slope of the diagram we can calculate the angular acceleration

We repeated the measurements adding a soft, deforming surface on the inclined plane (specifically, a layer of polystyrene). Our measurements are depicted in Fig. 23.3.

The angle of the incline plane was $7°$, so the angular acceleration as calculated from the trend line of the diagram is $a_r = 5.68$ rad/s^2. There is a deviation of the theoretical value of radial acceleration of about 75% due to the deformation of the surface and the ensuing rolling friction.

The proposed experimental setup provides the opportunity to obtain direct measurements of the angular velocity of a rigid body and, therefore, to calculate the moment of inertia.

We have chosen to study the measurement of the moment of inertia, among other given possibilities, because it is a mandatory exercise for schoolchildren in Greece. It is usually proposed that these measurements be made by means of photogates or chronometers. Our method allows schoolchildren to make such measurements with smartphones.

References

1. Monteiro, M., Cabeza, C., Marti, A.C.: Rotational energy in a physical pendulum. Phys. Teach. **52**, 181 (2014)
2. Monteiro, M., Cabeza, C., Marti, A.C., Vogt, P., Kuhn, J.: Angular velocity and centripetal acceleration relationship. Phys. Teach. **52**, 313 (2014)
3. Vogt, P., Kuhn, J.: Analyzing free fall with a smartphone acceleration sensor. Phys. Teach. **50**, 182 (2012)
4. Chevrier, J., Madani, L., Ledenmat, S., Bsiesy, A.: Teaching classical mechanics using smartphones. Phys. Teach. **51**, 376 (2013)
5. Shakur, A., Sinatra, T.: Angular momentum. Phys. Teach. **51**, 564 (2013)
6. Kuhn, J., Vogt, P.: Analyzing spring pendulum phenomena with a smartphone acceleration sensor. Phys. Teach. **50**, 504 (2012)

7. Vogt, P., Kuhn, J.: Analyzing collision processes with the smartphone acceleration sensor. Phys. Teach. **52**, 118–119 (2014)
8. Vogt, P., Kuhn, J.: Analyzing radial acceleration with a smartphone acceleration sensor. Phys. Teach. **51**, 182–183 (2013)
9. Vogt, P., Kuhn, J.: Analyzing simple pendulum phenomena with a smartphone acceleration sensor. Phys. Teach. **50**, 439–440 (2012)
10. Monteiro, M., Cabeza, C., Martí, A.C.: Exploring phase space using smartphone acceleration and rotation sensors simultaneously. Eur. J. Phys. **35**, 045013 (2014)
11. Hochberg, K., Gröber, S., Kuhn, J., Müller, A.: The spinning disc: studying radial acceleration and its damping process with smartphones' acceleration sensor. Phys. Educ. **49**(2), 137–140 (2014)
12. Silva, N.: Magnetic field sensor. Phys. Teach. **50**, 372 (2012)
13. Forinash, K., Wisman, R.F.: Smartphones as portable oscilloscopes for physics labs. Phys. Teach. **50**, 242 (2012)
14. Thoms, L.-J., Colicchia, G., Girwidz, R.: Color reproduction with a smartphone. Phys. Teach. **51**, 440 (2013)
15. Castro-Palacio, J.C., Velazquez-Abad, L., Gimenez, M.H., Monsoriu, J.A.: Using a mobile phone acceleration sensor in physics experiments on free and damped harmonic oscillations. Am. J. Phys. **81**, 472–475 (2013)
16. Sans, J., Manjón, F., Pereira, A., Gomez-Tejedor, J., Monsoriu, J.: Oscillations studied with the smartphone ambient light sensor. Eur. J. Phys. **34**, 1349 (2013)
17. Parolin, S.O., Pezzi, G.: Smartphone-aided measurements of the speed of sound in different gaseous mixtures. Phys. Teach. **51**, 508 (2013)
18. Kuhn, J., Vogt, P.: Analyzing acoustic phenomena with a smartphone microphone. Phys. Teach. **51**, 118 (2013)
19. https://ogy.de/physicstoolboxgyroscope

Part V
Mechanics of Deformable Bodies

Surface Tension Measurements with a Smartphone

Surface Tension Measurements with a Smartphone

24

Nicolas-Alexandre Goy, Zakari Denis, Maxime Lavaud,
Adrian Grolleau, Nicolas Dufour, Antoine Deblais,
and Ulysse Delabre

Smartphones are increasingly used in higher education and at university in mechanics (Chaps. 6 and 29) [1–3], acoustics (Chap. 44) [4], and even thermodynamics [5] as they offer a unique way to do simple science experiments. In this chapter we show how smartphones can be used in fluid mechanics to measure surface tension of various liquids, which could help students understand the concept of surface tension through simple experiments.

24.1 Background

Interfacial tension is the energy per unit area required for a material to create an interface with a surrounding material. This surface tension exists for interfaces between solids, liquids, and gas. In the case of a liquid surrounded by air, interfacial tension is often called surface tension, which plays an essential role in many natural phenomena such as pulmonary breathing, use of detergents, and insect-walking on water [6–8]. Usually surface tension is measured with quite expensive equipment (Wilhelmy plate, ring method) and long calibration procedures in order to get very accurate values, which perhaps prevents a simple understanding of the concept of surface tension. The method described here is based on the standard pendant drop method [9], where a drop of liquid is suspended from a tube as shown in Fig. 24.1a. The shape of the drop is then governed by the balance of surface tension forces along the tube and the weight of the drop. We show here that without a specific image setup, smartphones are accurate enough to get reasonable values of surface tension for various liquids such as water, oil, and water with detergent, and can be done easily by students on their smartphones.

N.-A. Goy (✉) · Z. Denis · M. Lavaud · A. Grolleau · N. Dufour · A. Deblais · U. Delabre
University of Bordeaux, CNRS, LOMA, UMR 5798, Talence, France

© The Author(s), under exclusive license to Springer Nature Switzerland AG 2022
J. Kuhn, P. Vogt (eds.), *Smartphones as Mobile Minilabs in Physics*,
https://doi.org/10.1007/978-3-030-94044-7_24

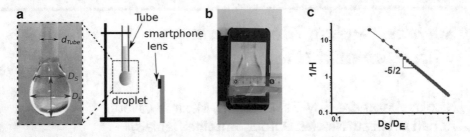

Fig. 24.1 (a) Water droplet suspended at the tip of a Pasteur pipette. (b) Illustration of the measurement of the diameters directly on the screen of the smartphone with a simple ruler. (c) $1/H$ as a function of the ratio D_S/D_E [10]. Note the logarithmic scale on the axes

When a drop is hung at the tip of a tube, the shape of the drop is governed by the local Laplace equation, where the weight of the drop and surface tension forces are balanced. The complete resolution of this equation requires computational [9, 10] analysis, which is the basis of the commercial measurement of surface tension.

However another way to estimate surface tension can be given by the following simple equation [10]:

$$\gamma = \frac{\Delta \rho g D_E^2}{H},\qquad(24.1)$$

where $\Delta \rho$ is the difference in density between the liquid and air, g is the gravitational constant ($g = 9.81$ m/s^2), D_E is defined in Fig. 24.1a and is the maximum diameter of the pendant drop, and

$$\frac{1}{H} = f\left(\frac{D_S}{D_E}\right)$$

is a dimensionless function of the ratio between D_S and D_E, which accounts for the specific shape of the drop due to gravity. D_S is defined as the diameter of the drop at a distance D_E from the bottom of the drop. For instance, if the drop was a perfect sphere, D_S would be zero. Due to gravity, the droplet is then elongated and D_S becomes strictly positive. This $1/H$ function can then be understood as a form factor and has been calculated numerically [9], but for our purpose we use the values tabulated in classic books [10] and represented in Fig. 24.1c. However, it is also interesting to note that within a good approximation, this $1/H$ dimensionless function can be approximated by a simple analytical formula,

$$\frac{1}{H} = a\left(\frac{D_S}{D_E}\right)^b,$$

with $a \approx 0.345$ and $b \approx -2.5$. Thus, following Eq. (24.1), surface tension can be obtained by measuring diameters with a ruler directly on the screen of the smartphone using the optical and numerical zooms of the smartphone.

24.2   Experiment

Droplets of various liquids (water, olive oil, water with surfactants, ethanol) were suspended at the tip of a Pasteur pipette (diameter tip $d_{\text{Tube}} = 1.4$ mm) by dipping it and pulling it out of a liquid reservoir. The experiment has also been tested with standard straws such as the tube of a pen ($d_{\text{Tube}} \approx 2.9$ mm) or McDonald's straw ($d_{\text{Tube}} \approx 5.9$ mm) instead of Pasteur pipette to test the generality of our approach. The smartphone used in the experiment is an iPhone 4 s (camera sensor 8 MPixels, display 640×960 pixels, 3.5-in screen) and the standard camera function of the smartphone is used. Taking advantage of the optical lens and the digital zoom ($\times 5$) of the smartphone, it is possible to measure D_{E}, D_{S}, and d_{Tube} directly on the screen of the smartphone with a standard ruler. For example, a droplet with a real diameter $D_{\text{E}} = 3.3$ mm (Fig. 24.1a) can be magnified using the optics and the digital zoom of the smartphone into a 2.8-cm diameter, which can easily be measured. For more accurate measurements, it is also possible to export pictures and analyze them with the free image analysis software ImageJ [11].

24.3   Results

Figure 24.2 shows typical pictures of various droplets at room temperature (20 °C) taken with the smartphone. Using Eq. (24.1) we get surface tension values summarized in Table 24.1. These results show that relatively good surface tension values for various liquids are obtained with this smartphone method. Even if the values obtained for surface tension for water are scattered, the values are always larger than the surface tension of olive oil and ethanol droplets, which is consistent with literature data. Indeed, a rough estimate of surface tension is given by [12]

$$\gamma \approx \frac{\text{E}}{a^2},$$

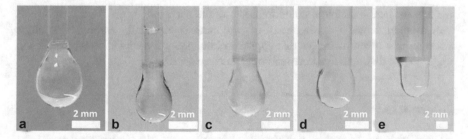

Fig. 24.2 (a) Water droplet, (b) oil droplet, (c) ethanol droplet hung at the tip of a Pasteur pipette. (d) Water droplet hung at the tip of a pen tube. (e) Water droplet hung at the tip of a McDonald's straw

Table 24.1 Surface tension values obtained with the smartphone pendant drop method

	Typical range of D_S/D_E	Density difference $\Delta\rho = \rho_L - \rho_{air}$ (kg/m^3)	Surface tension (mN/m) (Smartphone measurement at 20 °C)	Surface tension tabulated value (mN/m) at 20 °C
Water (Pasteur pipette)	0.68–0.78	999	78 ± 15 (6 drops)	72.8
Olive oil Pasteur pipette)	0.75	930	33 ± 3 (2 drops)	32
Ethanol Pasteur pipette)	0.8	780	22 ± 2 (2 drops)	23
Water (pen tube)	0.82	999	60 (2 drops)	72.8
Water McDonald's straw)	≈1	999	67	72.8

where E is the typical energy between two molecules and a the typical size of the molecules. This explains that the surface tension of water due to hydrogen bonds is much higher than that of olive oil or ethanol. For other measurements with standard straws, the experimental surface tension gives reasonable values but they are less accurate as explained in [13] because the diameter of the tube is quite large, especially compared to capillary length

$$L_c = \sqrt{\frac{\gamma}{\Delta\rho g}},$$

which induces D_S/D_E ratios close to 1 (see for example Fig. 24.2e).

24.4   Conclusion

We present here a simple method to measure the surface tension of various liquids with a smartphone. The simplicity of our method especially compared to force measurement methods makes this approach interesting for educational purposes. It could enable a simple characterization of surface tension.

References

1. Vogt, P., Kuhn, J.: Analyzing simple pendulum phenomena with a smartphone acceleration sensor. Phys. Teach. **50**, 439 (2012)
2. Vogt, P., Kuhn, J.: Analyzing free fall with a smartphone acceleration sensor. Phys. Teach. **50**, 182 (2012)

3. Madani, L., Ledenmat, S., Bsiesy, A., Chevrier, J.: Teaching classical mechanics using smartphones. Phys. Teach. **51**, 376 (2013)
4. Hirth, M., Kuhn, J., Müller, A.: Measurement of sound velocity made easy using harmonic resonant frequencies with everyday mobile technology. Phys. Teach. **53**, 120 (2015)
5. Elizabeth Vieyra, R., Vieyra, C., Macchia, S.: Kitchen physics: Lessons in fluid pressure and error analysis. Phys. Teach. **55**, 87 (2017)
6. Hu, D.L., Bush, J.W.M.: Meniscus climbing insects. Nature. **437**, 733–736 (2005)
7. Mohamed Boutinguiza Larosi: Floating together on the top. Phys. Teach. **53**, 93 (2015)
8. Ondris-Crawford, R.J., Hilliard, L.: It's all on the surface. Phys. Teach. **35**, 100 (1997)
9. Stauffer, C.E.: The measurement of surface tension by the pendant drop technique. J. Phys. Chem. **69**, 1933 (1965)
10. Adamson, A.W.: Physical Chemistry of Surfaces, 2nd edn. Interscience Publishers (1967)
11. ImageJ software. https://imagej.nih.gov/ij/download.html
12. Berry, M.V.: The molecular mechanism of surface tension. Phys. Educ. **6**(2), 79–84 (1971)
13. Berry, J.D., Neeson, M.J., Dagastine, R.R., Chan, D.Y., Tabor, R.F.: Measurement of surface and interfacial tension using pendant drop tensiometry. J. Colloid. Interface Sci. **454**, 226 (2015)

Exploring the Atmosphere Using Smartphones

25

Martín Monteiro, Patrik Vogt, Cecilia Stari, Cecilia Cabeza, and Arturo C. Marti

The characteristics of the inner layer of the atmosphere, the troposphere, are determinant for Earth's life. In this experience we explore the first hundreds of meters using a smartphone mounted on a quadcopter. Both the altitude and the pressure are obtained using the smartphone's sensors. We complement these measures with data collected from the flight information system of an aircraft. The experimental results are compared with the International Standard Atmosphere and other simple approximations: isothermal and constant density atmospheres.

25.1 The International Standard Atmosphere

The atmospheric conditions exhibit strong variations at different points and different times. To provide a unified frame of reference, an atmospheric model, the International Standard Atmosphere (ISA), has been established [1]. It consists of tables of pressure, temperature, density, and other variables suitable at mid-latitudes over a wide range of altitudes. The ISA is used for several purposes ranging from altimeter calibration to comparison of aircraft performance among others. The ISA is divided into layers with simple temperature variations, depicted in Fig. 25.1. The inner layer is the troposphere, from the surface until 11 km of height, in which the temperature presents a linear gradient, named lapse rate, $C = -0.0065\ °C/m$.

M. Monteiro (✉)
Universidad ORT Uruguay, Montevideo, Uruguay
e-mail: monteiro@ort.edu.uy

P. Vogt
Institute of Teacher Training (ILF) Mainz, Mainz, Germany
e-mail: vogt@ilf.bildung-rp.de

C. Stari · C. Cabeza · A. C. Marti
Universidad de la República, Montevideo, Uruguay
e-mail: cstari@fing.edu.uy; cecilia@fisica.edu.uy; marti@fisica.edu.uy

© The Author(s), under exclusive license to Springer Nature Switzerland AG 2022
J. Kuhn, P. Vogt (eds.), *Smartphones as Mobile Minilabs in Physics*,
https://doi.org/10.1007/978-3-030-94044-7_25

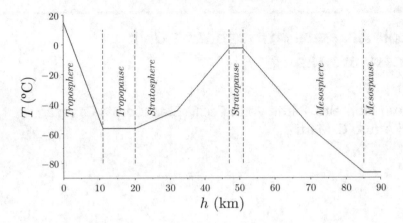

Fig. 25.1 International Standard Atmosphere

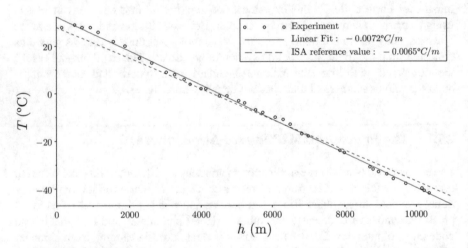

Fig. 25.2 Temperature as a function of the altitude using an aircraft's information system. The reference value of the temperature lapse rate in the troposphere according to the ISA is also plotted

The flight information system usually available to the passengers in many aircraft provides a clever way to quantify the relationship between temperature and pressure. In Fig. 25.2 we plot the temperature as a function of the altitude using data collected by a passenger. The linear fit gives a temperature gradient of about $C = -0.0072$ °C/m. Of course, this value highly depends on the particular atmospheric conditions during this flight and does not necessarily represent accurately the ISA. This exercise can be proposed as homework to students about to travel on a plane.

In addition to the ISA we consider two further atmospheric models. A rather crude approximation consists of considering the temperature constant in the inner layers, in contrast to the model considered above. This model is called isothermal

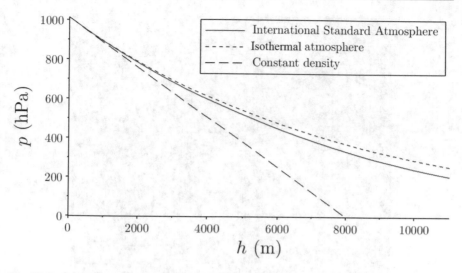

Fig. 25.3 Pressure as a function of the altitude for the three different models of the atmosphere considered: the International Standard Atmosphere (ISA), the isothermal, and the constant density atmosphere

atmosphere. Using the ideal gas law, an exponential expression for the pressure as a function of the altitude is obtained [2].

Finally, another approximation is obtained neglecting air density variations. Under this approximation valid at small altitudes (a few kilometers), ρ is constant and the atmospheric pressure obeys the hydrostatic equation

$$p = p_0 - \rho g h,$$

where p_0 is the pressure at $z = 0$. For the sake of comparison, the pressure p as a function of the height h is shown in Fig. 25.3 for the three different models considered here.

25.2 The Experiment

In the present experiment we use the pressure sensor and the GPS of a smartphone attached to a quadcopter. Other interesting experiments involving pressure sensors or GPS can be found in [3, 4] (Chap. 42), and the use of quadcopters in teaching physics has been recently considered in [5].

A smartphone LG model G3 was mounted in a DJI Phantom 2 using an armband case as seen in Fig. 25.4. The quadcopter was raised to a height of 250 m with respect to the takeoff point, kept hovering for a dozen seconds, and taken down. During the flight, the Androsensor app was used to register the atmospheric pressure using the built-in barometer and the height obtained from the GPS. As the response time of the GPS is rather slow, during ascending and descending the operator tried to control

Fig. 25.4 Smartphone mounted on a DJI Phantom 2 using an armband

the thrust of the device in such a way that the vertical speed was roughly constant. According to the information provided by the manufacturer of the quadcopter, the maximum ascent speed is 6 m/s and the maximum descent speed is 2 m/s. Care should be taken to avoid damage to persons, animals, or property and also to fulfill air traffic regulations [6].

25.3 Results

In Fig. 25.5, the pressure as a function of the altitude is shown. From the slope of the linear fit the air density is found to be $\rho = 1.12$ kg/m^3. This value is a good approximation of the standard value valid in the inner atmospheric layer. In the conditions of the present experiment the standard value of density is 1.18 kg/m^3.

25.4 Conclusion

The use of a quadcopter and a smartphone allowed us to record the main atmospheric variables. In this experiment, we obtained pressure profiles and, through linear fits, values of the density and compared them with standard values available in the literature. To gain further insight into the characteristics of the inner layers of the atmosphere, observations of a passenger in a commercial aircraft are also reported.

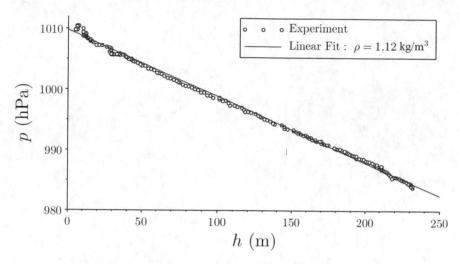

Fig. 25.5 Atmospheric pressure as a function of the altitude recorded by the smartphone

This experiment helps to get an insight into the characteristics of the atmosphere using accessible tools.

References

1. U.S. Standard Atmosphere, 1976. NASA. Available at http://ntrs.nasa.gov/archive/nasa/casi.ntrs. nasa.gov/19770009539.pdf or more updated information at https://en.wikipedia.org/wiki/ International_Standard_Atmosphere
2. Lee, K.M., Ryan, D.M.: Scale height—a parameter for characterizing atmosphere. Phys. Teach. **53**, 122–123 (2015)
3. Gabriel, P., Backhaus, U.: Kinematics with the assistance of smartphones: measuring data via GPS-Visualizing data with Google Earth. Phys. Teach. **51**, 246–247 (2013)
4. Müller, A., Hirth, M., Kuhn, J.: Tunnel pressure waves – A mobile phone inquiry on a rail travel. Phys. Teach. **54**, 118–119 (2016)
5. Monteiro, M., Stari, C., Cabeza, C., Marti, A.C.: Analyzing the flight of a quadcopter using a smartphone. http://arxiv.org/abs/1511.05916 (2015)
6. Juniper, A.: *The Complete Guide to Drones*. Hachette (2015)

On the Inflation of a Rubber Balloon

26

Julien Vandermarlière

It is a well-known fact that it is difficult to start a balloon inflating. But after a pressure peak that occurs initially, it becomes far easier to do it! The purpose of this chapter is to establish the experimental pressure-radius chart for a rubber balloon and to compare it to the theoretical one. We will demonstrate that the barometer of a smartphone is a very suitable tool to reach this goal. We hope that this phenomenon will help students realize that sometimes very simple questions can lead to very interesting and counterintuitive science.

26.1 Theoretical Background

Rubber balloons have been very good toys for children for decades. And for physicists too! Indeed, they do not deform according to Hook's law, but rather they expand in a strange way as we shall see below. They are made of long flexible chains of polymer that are coiled and joined together by randomly oriented chemicals bonds [1] (Fig. 26.1).

When we inflate a balloon, those chains start to be straightened out. The average distance between bonds increases, and this initiates the restoring force, which tends to return the balloon to its initial shape. For a short time, it is very difficult to fight against it. But once the chains are stretched a little bit, they become more organized and the rubber becomes more stretchy. The pressure inside the balloon now evolves like the pressure inside a bubble soap: it decreases as the radius increases. But at large elongation the chains are completely stretched out and new phenomena like strain-induced crystallization result in some stiffening; the pressure goes up again!

J. Vandermarlière (✉)
Lycée Jean Lurçat, Perpignan, France
e-mail: julien.vandermarliere@ac-montpellier.fr

Fig. 26.1 Chains of polymer in a rubber material at different elongation

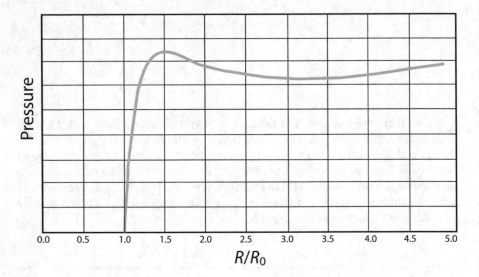

Fig. 26.2 Theoretical pressure-radius curve [see Eq. (26.1)]

A number of physicists have been interested in this behavior and introduced a variety of physical descriptions [1–3]. Verron and Marckmann [3] proposed Eq. (26.1) in the case of a Mooney-Rivlin type spherical balloon:

$$P = K \left[\frac{R_0}{R} - \left(\frac{R_0}{R} \right)^7 \right] \left[1 + 0.1 \left(\frac{R}{R_0} \right)^2 \right], \tag{26.1}$$

where R_0 is the initial radius and R the deformed radius. K is a coefficient dependent on the material parameters. The associated inflation curve looks like Fig. 26.2.

26.2   The Experiment

It has been reported in this column that smartphone barometers are good enough for real physics experiments (Chap. 4) [4], so a Samsung S5 was inserted into a balloon. Hopefully, the rubber is a tactile enough material and is transparent enough to see the screen so that we can still control the smartphone!

Fig. 26.3 Screenshot of the
Barometer Graph app
(pressure is in hPa, time in
seconds is on the horizontal
axis)

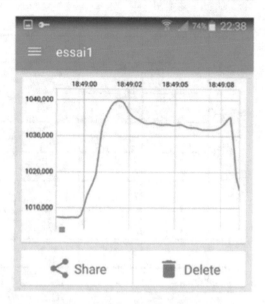

Fig. 26.4 The smartphone in a balloon! The Physics Toolbox app is controlled from the computer
via Wi-Fi with the SideSync app

The Barometer Graph [5] app was launched, and the evolution of the
pressure vs. time was recorded while the balloon was being inflated. A single
exhalation was used to inflate the balloon, made as regular as possible. A screenshot
of the result is shown in Fig. 26.3.

First, the pressure increases quickly and reaches a peak. Then it decreases, which
is very counterintuitive! At the end of the inflation the pressure goes up again.
Finally, the inflator's lips and balloon are separated and the balloon deflates. The
main characteristics of the theory are seen.

In order to obtain better measurements, a second experiment was performed. This
time the Physics Toolbox Sensor Suite app was used to measure the pressure [6]. The
smartphone was controlled via Wi-Fi from a computer using the SideSync app [7]
(Fig. 26.4). While the balloon was inflated, the circumference was measured as well.

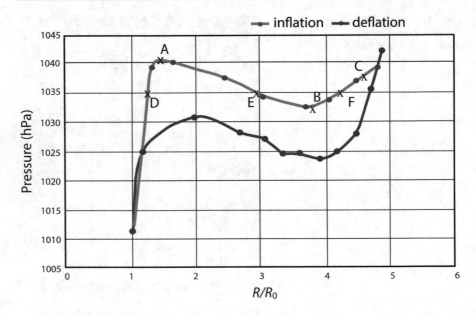

Fig. 26.5 Experimental chart

Then a spreadsheet was used to plot the pressure as a function of the radius ratio, $P = f(R/R_0)$ (Fig. 26.5).

Once again the experimental inflation curve (in blue) matches the theoretical curve. The pressure peak (1040.7 hPa) is only 3% greater than the atmospheric pressure and occurs when R/R_0 is about 1.44 (point A). This matches fairly well the theoretical prediction of Verron and Marckmann [3], which is 1.48 [from Eq. (26.1)]. The difference is only 2.8%. We also looked at the deflation of the balloon (in red). An interesting hysteresis phenomenon is clearly visible. It can be explained by the fact that when the balloon returns to its original shape, the polymer chains never get back to their initial orientations. This measurement process was repeated with several balloons and the experiment consistently produced similar results.

26.3 Comments

This work can help to explain what seems to be a paradox: the two balloons experiment [8]. If you connect two balloons with a hollow tube, one big and one smaller, and let the air rush between them, the small one can inflate the big one! This is because most of the time the balloons are on the [AB] interval of the chart in Fig. 26.5. But the opposite is also possible if the smallest balloon is at B and the biggest at C. It is possible to reach equilibrium as well. Take three balloons and inflate them to reach the D, E, and F points. Connect them together and... nothing will happen. Their internal pressures are exactly the same! This experiment can easily be done in the classroom, and students may learn lots of interesting things while doing it!

References

1. Stein, R.S.: On the inflating of balloons. J. Chem. Educ. **35**(4), 203 (1958)
2. Merritt, D.R., Weinhaus, F.: The pressure curve for a rubber balloon. Am. J. Phys. **46**, 976–978 (1978)
3. Verron, E., Marckmann, G.: Numerical analysis of rubber balloons. Thin-Walled Structures. **41**(8), 731–746 (2003)
4. Müller, A., Hirth, M., Kuhn, J.: Tunnel pressure waves – A mobile phone inquiry on a rail travel. Phys. Teach. **54**, 118–119 (2016)
5. https://m.apkpure.com/de/barometer-graph/com.ghsoft.barometergraph
6. https://ogy.de/physicstoolboxsensorsuite
7. https://ogy.de/sidesyncandroid
8. Weinhaus, F., Barker, W.: On the equilibrium states of interconnected bubbles or balloons. Am. J. Phys. **46**, 978 (1978)

Video Analysis on Tablet Computers to Investigate Effects of Air Resistance

27

Sebastian Becker-Genschow, Pascal Klein, and Jochen Kuhn

In recent years several approaches have been published using video analysis in physics education in an innovative way [1–4]. The common idea is that students record the motion, transfer the video on a computer, and analyze it with suitable software. In this chapter we demonstrate how students can perform video analysis on mobile devices such as smartphones and tablet computers by using special applications. By doing so, students can record and analyze the motion during the experimental process and combine the advantages of mobile devices with the possibilities of video motion analysis [5, 6]. Moreover the possibility of recording slow motion videos allows the analysis of even fast-moving objects. In our presented approach we utilize video analysis on tablet computers in an attractive experimental setup for students to investigate the effects of air resistance on falling objects with low-cost material by themselves.

27.1 Theoretical Background

For typical classroom experiments with falling objects in air, we find large Reynolds numbers $R \gg 1$ [7]. For that reason we can assume that the magnitude of the drag force is proportional to the square of the fall velocity and can therefore be written as

S. Becker-Genschow (✉)
Digitale Bildung, Department Didaktiken der Mathematik und der Naturwissenschaften, Universität zu Köln, Cologne, Germany
e-mail: sebastian.becker-genschow@uni-koeln.de; sbeckerg@uni-koeln.de

P. Klein
Georg-August-University Göttingen, Göttingen, Germany
e-mail: pklein@physik.uni-kl.de

J. Kuhn
Ludwig-Maximilians-Universität München (LMU Munich), Faculty of Physics, Chair of Physics Education, Munich, Germany
e-mail: jochen.kuhn@lmu.de

$$F = k \cdot v^2, \tag{27.1}$$

where the proportionality constant is given by

$$k = 0.5 \cdot C_D \cdot \rho \cdot A. \tag{27.2}$$

Here ρ denotes the density of air, A the projected surface area of the falling object, and C_D the drag coefficient. If the drag force is equal to gravitational force, the falling object reaches its terminal velocity v_t, given by

$$v_t = \sqrt{\frac{2 \cdot m \cdot g}{C_D \cdot \rho \cdot A}}, \tag{27.3}$$

where m denotes the mass of the falling object. Note (i) that the square of terminal velocity is linearly dependent on the mass of the object,

$$v_t^2 = b \cdot m. \tag{27.4}$$

with

$$b = \frac{2 \cdot g}{C_D \cdot \rho \cdot A}, \tag{27.5}$$

and note (ii) that the drag coefficient can be expressed as

$$C_D = \frac{2 \cdot g}{b \cdot \rho \cdot A}. \tag{27.6}$$

27.2   Experimental Setup

Since students should be able to perform measurements in school labs, we have to choose an object that reaches its terminal velocity before impact. Therefore we adopted the idea of using low-cost paper cups (Fig. 27.1, right) for fall experiments [8]. To determine the terminal velocity, one student drops the paper cup and a second student records the motion with a tablet computer perpendicular to the plane of motion from a sufficient distance (approximately 3 m). Note that a yardstick is placed in the picture to aid in converting pixel distances to real distances (Fig. 27.1, left). The video is recorded with a minimum camera frequency rate of 60 frames per second (fps) in order to prevent cross-fading. For recording and analyzing the motion of the falling object, we used iPad minis and the free application Viana [9].

Fig. 27.1 Small paper cup used for fall experiments (right) and experimental setup (left)

27.3   Experimental Results

Determination of Terminal Velocity

Video analysis yields the position of the paper cup and its velocity at points of time predefined by the camera frequency rate. From the time-velocity graph students can easily conclude that the velocity of fall converges to a terminal velocity (see Fig. 27.2).

However, more accurate values can be obtained from the time-position data. Therefore, the measured data can be transferred from Viana to the free application Vernier Graphical Analysis [10] to perform a linear regression analysis in the section where the velocity of fall reaches the terminal velocity (Fig. 27.3) and the fall velocity can be identified as the slope of the regression curve (here: $v_t = 1.425$ m/s with RMSE = 0.004).

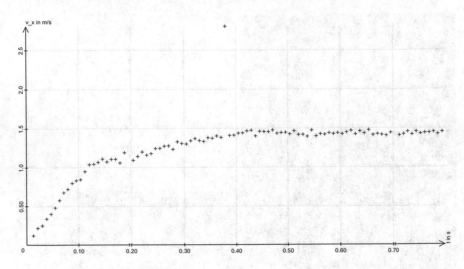

Fig. 27.2 Screenshot of the time-velocity diagram provided by Viana [9]

Determination of the Drag Coefficient

The mass of the falling object can be doubled, tripled, etc. by simply stacking the paper cups. By doing this only the mass is changed whereas the projected surface area as well as the drag coefficient does not change, thus allowing for ready determination of the terminal velocities for one to five paper cups experimentally. The terminal velocity square vs. mass was plotted and a least-squares model fit using Vernier Graphical Analysis [10] was performed, which confirmed the theoretically expected behavior according to Eq. (27.4) (Fig. 27.4).

We used Eq. (27.6) to obtain the drag coefficient from the fit parameter. By inserting the following values: $m = 0.34$ g, $\rho = 1.20$ kg/m^3, and $A = 44.18$ cm^2 we calculated the drag coefficient to $C_D = 0.59$. If we compare this calculated value with the drag coefficient of a sphere ($C_D = 0.47$) [11] it is larger, as one might expect since for the same projected surface area a sphere has less aerodynamic drag because of its shape.

27.4 Conclusions

We have presented an experimental approach for students to explore effects of air resistance by performing video analysis on tablet computers. With our experimental setup students can identify the asymptotic behavior of the fall velocity and determine the terminal velocity as well as the drag coefficient by only using a tablet computer equipped with special applications. Using tablet computers as measurement devices enables students to investigate physical relationships with a high degree of auton- omy with little experimental effort and compare the experimental data with

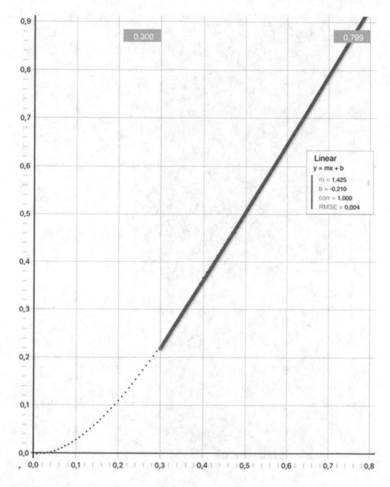

Fig. 27.3 Screenshot of linear regression analysis using Vernier Graphical Analysis [10]

theoretical relations just-in-time on the same device. This theory-experiment inter-play is an important method for learning physics [12]. Since the use of mobile devices is becoming more common in introductory and advanced physics courses, our approach could serve as a basis for motivating classroom activities for students, too. More advanced students might benefit from going beyond the concept of terminal velocity to the full-fledged solution for the equation of motion [13].

Learning with mobile video analysis can also increase conceptual understanding [14, 16] while decreasing irrelevant cognitive effort and negative emotions [15, 16].

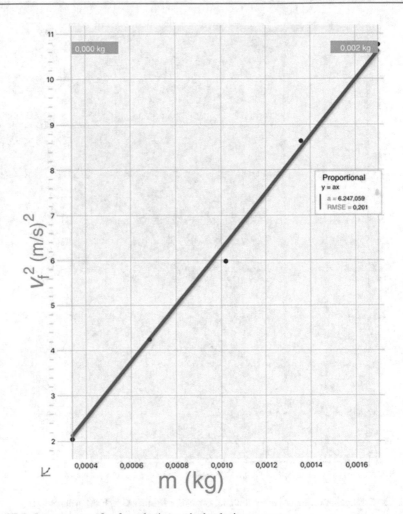

Fig. 27.4 Least-squares fit of quadratic terminal velocity vs. mass

References

1. Brown, D., Cox, A.: Innovative uses of video analysis. Phys. Teach. **47**, 145 (2009)
2. Desbien, D.: High-speed video analysis in a conceptual physics class. Phys. Teach. **49**, 332 (2011)
3. Wyrembeck, E.: Video analysis with a web camera. Phys. Teach. **47**, 28 (2009)
4. Klein, P., Gröber, S., Kuhn, J., Fleischhauer, A., Müller, A.: The right frame of reference makes it simple: an example of introductory mechanics supported by video analysis of motion. Eur. J. Phys. **36**(1), 015004 (2015)
5. Kuhn, J.: Relevant information about using a mobile phone acceleration sensor in physics experiments. Am. J. Phys. **82**, 94 (2014)

6. Klein, P., Gröber, S., Kuhn, J., Müller, A.: Video analysis of projectile motion using tablet computers as experimental tools. Phys. Educ. **49**(1), 37–40 (2013)
7. Takahashi, K., Thompson, D.: Measuring air resistance in a computerized laboratory. Am. J. Phys. **67**, 709 (1999)
8. Angell, C., Ekern, T.: Measuring friction on falling muffin cups. Phys. Teach. **37**, 181 (1999)
9. Viana requires iOS 8.1 or higher. This application is available from https://goo.gl/4RWv8g
10. Vernier Graphical Analysis requires iOS 8.0 or higher or Android 4.0 or higher. This application is available from https://goo.gl/BjNtNg (for iOS) and https://goo.gl/za5euQ (for Android)
11. http://arc.id.au/CannonballDrag.html
12. Gröber, S., Klein, P., Kuhn, J.: Video-based problems in introductory mechanics physics courses. Eur. J. Phys. **35**(5), 055019 (2014)
13. Cross, R., Lindsey, C.: Measuring the drag force on a falling ball. Phys. Teach. **52**, 169 (2014)
14. Becker, S., Gößling, A., Klein, P., Kuhn, J.: Using mobile devices to enhance inquiry-based learning processes. Learn. Instr. **69**, 101350 (2020)
15. Hochberg, K., Becker, S., Louis, M., Klein, P., Kuhn, J.: Using smartphones as experimental tools—a follow-up: cognitive effects by video analysis and reduction of cognitive load by multiple representations. J. Sci. Educ. Technol. **29**(2), 303–317 (2020)
16. Becker, S., Gößling, A., Klein, P., Kuhn, J.: Investigating dynamic visualizations of multiple representations using mobile video analysis in physics lessons: effects on emotion, cognitive load and conceptual understanding. Zeitschrift für Didaktik der Naturwissenschaften. **26**(1), 123–142 (2020)

Determination of the Drag Resistance Coefficients of Different Vehicles

28

Christoph Fahsl and Patrik Vogt

While it has been demonstrated how air resistance could be analyzed by using mobile devices (Chap. 27) [1], this chapter demonstrates a method of how to determine the drag resistance coefficient c of a commercial automobile by using the acceleration sensor of a smartphone or tablet. In an academic context, the drag resistance is often mentioned, but little attention is paid to quantitative measurements. This experiment was driven by the fact that this physical value is most certainly neglected because of its difficult measurability. In addition to that, this experiment gives insights on how the aerodynamic factor of an automobile affects the energy dissipation and thus how much power is required by automobile transportation.

Other works describe efforts to determine the drag resistance coefficient of cars, wherein the speed profile during the rolling process is determined by reading the speedometer [2], using logged GPS data [3], or by analyzing conservation of energy [4]. Apart from the GPS measurement, low material costs are advantages of these variants. However, a disadvantage is the small number of data points when reading the speedometer and the difficulty of calculating the velocity based on raw GPS data.

28.1 Theoretical Background

We are looking at a car that is decelerating due solely to the force of air drag (F_D) and the force of rolling drag (F_R) from a certain initial velocity. From Newton's second law we get:

C. Fahsl (✉)
University of Konstanz, Konstanz, Germany

P. Vogt
Institute of Teacher Training (ILF) Mainz, Mainz, Germany
e-mail: vogt@ilf.bildung-rp.de

© The Author(s), under exclusive license to Springer Nature Switzerland AG 2022
J. Kuhn, P. Vogt (eds.), *Smartphones as Mobile Minilabs in Physics*,
https://doi.org/10.1007/978-3-030-94044-7_28

Fig. 28.1 Front view of the car. Comparing the red and white areas allows one to determine the cross-sectional area

$$M \cdot a = F_D + F_R = \frac{1}{2} c\rho A v^2 + \mu M g \tag{28.1}$$

$$a = \frac{1}{2} \frac{c\rho A}{M} v^2 + \mu g, \tag{28.2}$$

where A is the effective cross-sectional area [$A = (2.22 \pm 0.05)$ m^2 (Fig. 28.1)], c is the drag resistance coefficient, μ is the rolling resistance coefficient, M is the mass of the car [$M = (1430 \pm 5)$ kg], v is the velocity, ρ is the air density [$\rho = (1.20 \pm 0.05)$ kg/m^3], and g is the gravity of Earth ($g = 9.81$ ms^{-2}).

By plotting $|a|$ vs. v^2, you get—according to the formula—a straight line. The slope $m = (c\rho A)/(2 M)$ provides the drag resistance coefficient.

28.2 Execution of the Experiment

For the execution of this experiment, the smartphone was attached to the car in a horizontal orientation. It's important that one acceleration sensor aligns in the direction of motion (Fig. 28.2).

Since the car has to be accelerated from a standstill position, a longer stretch of road with a parking place at the beginning is the best place to execute the experiment. Before driving on the road, the recording of the acceleration data must have already been started. After reaching a certain predetermined velocity, the driver puts the car in neutral and lets it roll until it reaches the desired velocity. After reaching that velocity, the measurement can be stopped. For general safety, another passenger should always be present that assists with the entire experiment. The driver must

Fig. 28.2 Mounting of the smartphone in the car and the orientation of one of the acceleration sensors in the direction of motion (here y sensor). The app used in this experiment is called Accelerometer Data Pro (Wavefront Labs, 2011). It is running on an iPhone 3G

never do the recording him/herself. Moreover, the experiment should be done on an uncrowded road in accordance with traffic laws.

28.3   Evaluation of Acceleration Data

The following graph shows the acceleration data the entire time of the experiment (Fig. 28.3). The first 30 s illustrate the process of acceleration. The large dips are due to shifting gears in the car. After that period, you can see the process of letting the car roll out until it reached its predetermined velocity.

We used the program Measure by PHYWE to smooth and integrate the data curves (alternatively, you can also use other software, e.g., PASCO Capstone.). By integrating the smoothed acceleration-data curve over time, you get the velocity curve of the car (Fig. 28.4).

The final step is now to plot $|a|$ vs. v^2 (Fig. 28.5) to get the characteristic line of Eq. (28.2). The resulting drag coefficients for three consecutive measurements are shown in Table 28.1.

The weighted average of those results is $c = 0.320 \pm 0.009$ (literature value in published sources [5] is 0.37). External influences like wind and uneven roads are most likely the reason for the deviation of the results in comparison to the literature value. But taking into account that only a smartphone was used in the whole process, we come to the conclusion that this method is a fast, efficient, and especially cheap

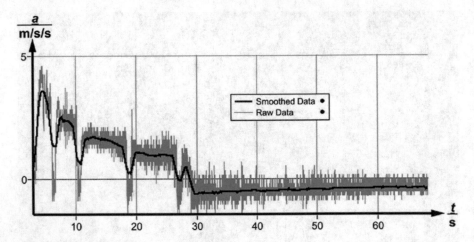

Fig. 28.3 The *a-t/v-t* diagram of the entire process

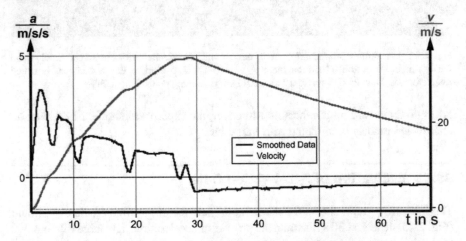

Fig. 28.4 The *a-t/v-t* diagram of the entire process

way (in comparison to the expensive use of a wind tunnel) to determine the characteristic drag coefficient of a car. In addition to the results shown above, the measurements were repeated with two other cars and a bicycle (Fig. 28.6).

In all cases we were able to reproduce results in agreement with literature values. The following tables show the results for a Volkswagen T3 bus, a fire engine, and a bicycle (Table 28.2). Again, we were able to reproduce results in agreement with literature values.

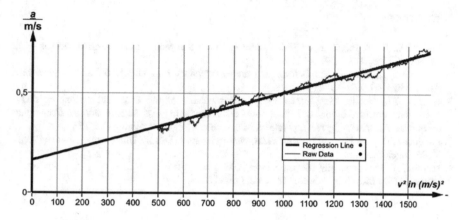

Fig. 28.5 Regression line of the acceleration data over v^2. Note that only data sampled from times after 30 s are plotted

Table 28.1 Slope of the regression line allows us to calculate the drag coefficients and their errors

Measurement	*slope* in m^{-1}	Δ *slope* in m^{-1}	c	Δc
1	0.000343	0.0000009	0.368	0.017
2	0.0003029	0.00000013	0.325	0.016
3	0.0002674	0.00000011	0.287	0.013

Fig. 28.6 Bus, fire engine, and bicycle used in additional measurements

Table 28.2 Repetition of the experiment with other cars and a bicycle

Value	VW T3 Bus	Fire engine	Bicyclist
Mass M	2000 $\pm$ 5 kg	12,442 kg	86.5 $\pm$ 0.1 kg
Effective cross-sectional area A	3.17 $\pm$ 0.05 m^2	4.90 $\pm$ 0.05 m^2	0.60 $\pm$ 0.05 m^2
Air density ρ	1.20 $\pm$ 0.05 kgm^{-3}	1.20 $\pm$ 0.05 kgm^{-3}	1.20 $\pm$ 0.05 kgm^{-3}
c (Literature values)	0.51 [5]	0.8–1.5 [6]	1.0 [7]
c (Experimental values)	0.501 $\pm$ 0.023	1.49 $\pm$ 0.07	1.12 $\pm$ 0.07

References

1. Becker, S., Klein, P., Kuhn, J.: Video analysis on tablet computers to investigate effects of air resistance. Phys. Teach. **54**, 440 (2016)
2. Kwasnoski, J., Murphy, R.: Determining the aerodynamic drag coefficient of an automobile. Am. J. Phys. **53**, 776 (1985)
3. Huertas, J.I., Coello, G.A.Á.: Accuracy and precision of the drag and rolling resistance coefficients obtained by on road coast down tests. Proceedings of the International Conference on Industrial Engineering and Operations Management Bogota, Colombia (2017)
4. Fieblinger, G.: Das Fahrrad im Unterricht. Naturwissenschaften im Unterricht. **1**, 8–11 (1983)
5. Volkswagen brochure.
6. Lindner, H.: Physik für Ingenieure. Hanser (2009)
7. Wilson, D.G.: Bicycling Science. The MIT Press (2010)

Part VI
Pendulums

Analyzing Simple Pendulum Phenomena with a Smartphone Acceleration Sensor

Patrik Vogt and Jochen Kuhn

This chapter describes a further experiment using the acceleration sensor of a smartphone. For a previous column on this topic, including the description of the operation and use of the acceleration sensor, see [1]. In this contribution we focus on analyzing simple pendulum phenomena. A smartphone is used as a pendulum bob, and SPARKvue [2] software is used in conjunction with an iPhone or an iPod touch, or the Accelogger [3] app for an Android device. As described in [1], the values measured by the smartphone are subsequently exported to a spreadsheet application (e.g., MS Excel) for analysis.

The following aspect should be emphasized before starting: If the lesson objective is to determine the period of an oscillation in an experiment, it is advisable to use a conventional stopwatch rather than a smartphone. The use of a cell phone as a pendulum bob, however, generates a much greater quantity of information and learning opportunities than the duration of a period alone and can greatly enhance instruction.

29.1 Investigation of the Mathematical Pendulum Using a Smartphone

In order to perform an experiment that examines the existing laws governing a simple pendulum, a smartphone can be suspended on two strings—this prevents rotation around the longitudinal axis (Fig. 29.1). Figure 29.2 shows a measurement example for a pendulum length $l = 1.15$ m (perpendicular distance between the

P. Vogt (✉)
Institute of Teacher Training (ILF) Mainz, Mainz, Germany
e-mail: vogt@ilf.bildung-rp.de

J. Kuhn
Ludwig-Maximilians-Universität München (LMU Munich), Faculty of Physics, Chair of Physics Education, Munich, Germany
e-mail: jochen.kuhn@lmu.de

Fig. 29.1 iPod touch
suspended from two strings
(shown in red)

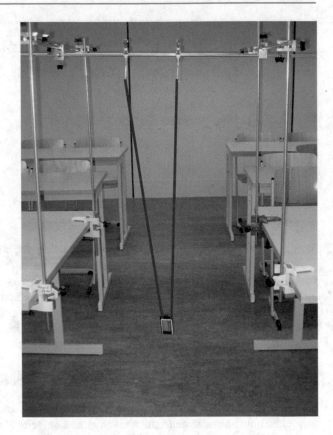

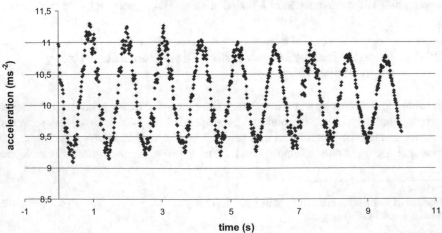

Fig. 29.2 Acceleration process for a mathematical pendulum ($l = 1.15$ m); presentation of
measured values after the export of data from the smartphone into MS Excel

center of mass of the iPod touch and the pivot point), and acceleration taking effect in the direction of the string is represented. The values measured constitute a basis for discussion on the following:

(a) The accelerations measured in the direction of the string (Fig. 29.2). Why do the acceleration values vary around the acceleration of gravity and at what amplitudes are minimum and maximum displacement reached?
(b) Determine the period of a complete oscillation and compare it to the value theoretically expected.
(c) Convert the tangential accelerations measured into pendulum amplitudes α or x with the help of a spreadsheet, in order to create an α-t or x-t diagram, which is easier for pupils to interpret.
(d) Discuss damping and calculate the logarithmic decrement.

At this point, items (a) and (b) can be examined in more detail: Regarding (a), the acceleration in the direction of the string is given by the sum of the centrifugal force apparently taking effect and the pendulum's mass in the direction of the string. As velocity is briefly zero at the turning point, the pendulum's mass only takes effect in the direction of the string, so the acceleration of small g is measured. The minima are thus located at the turning points on the acceleration curve (Fig. 29.2). When passing through the rest point, the pendulum is moving at its highest velocity and, as a result, at maximum centrifugal acceleration. In addition, the acceleration of gravity needs to be completely added to this, i.e., the accelerations at the zero crossing point are higher than g and correspond to the maxima in Fig. 29.2. This observation makes it clear that the time lag between two peaks corresponds to half a period.

Regarding (b), the time lag between the first and ninth maximum (Fig. 29.2) is 8.61 s and corresponds to the time required for four complete swings. The period of the string pendulum's swing determined in the experiment is therefore 2.153 s with a measurement error of ± 0.002 s. The duration of a complete swing T theoretically expected for a pendulum with a length $l = 1.15$ m is calculated with the formula

$$T = 2\pi\sqrt{\frac{l}{g}} \qquad (29.1)$$

to be 2.15 s and therefore matches well with the result of the experiment. Similar to the analysis of the free fall at a free-fall tower (Chap. 6) [1], it is interesting to examine the simple string pendulum and its oscillation process using an everyday object such as a playground swing. The particular appeal of this experiment is that the pupils can start by experiencing the acceleration with their own bodies. They can then perform an assessment of the oscillation process, which subsequently can be quantitatively tested. In order to carry out the experiment, the smartphone is fixed to the swing with adhesive tape, with one axis pointing in the direction of tangential acceleration and the other axis in the direction of radial acceleration (Fig. 29.3).

Fig. 29.3 Examination of the mathematical pendulum using a playground swing; the iPod touch is fixed to the swing with adhesive tape

Studying this phenomenon using smartphones' acceleration sensor should also be integrated in a more sophisticated instructional setting, connected with other phenomena [4–6] addressing the students' misconceptions which could be related to this concept [7]. Anyway, current research shows that using this method to study free fall and oscillation phenomena could at least increase curiosity and motivation of the students when they learn with smartphones or tablets as experimental tools [8].

References

1. Vogt, P., Kuhn, J.: Analyzing free-fall with a smartphone acceleration sensor. Phys. Teach. **50**, 182–183 (2012)
2. https://ogy.de/sparkvue
3. https://ogy.de/accelogger
4. Kuhn, J.: Relevant information about using a mobile phone acceleration sensor in physics experiments. Am. J. Phys. **82**(2), 94 (2014)

5. Kuhn, J., Vogt, P., Müller, A.: Analyzing elevator oscillation with the smartphone acceleration sensors. Phys. Teach. **52**(1), 55–56 (2014)
6. Vogt, P., Kuhn, J.: Analyzing collision processes with the smartphone acceleration sensor. Phys. Teach. **52**(2), 118–119 (2014)
7. Hall, J.: More smartphone acceleration. Phys. Teach. **51**(1), 6 (2013)
8. Hochberg, K., Kuhn, J., Müller, A.: Using smartphones as experimental tools – effects on interest, curiosity and learning in physics education. J. Sci. Educ. Technol. **27**(5), 385–403 (2018)

Measurement of g Using a Magnetic Pendulum and a Smartphone Magnetometer

30

Unofre Pili, Renante Violanda, and Claude Ceniza

The internal sensors in smartphones for their advanced add-in functions have also paved the way for these gadgets becoming multifunctional tools in elementary experimental physics [1]. For instance, the acceleration sensor has been used to analyze free-falling motion (Chap. 6) [2] and to study the oscillations of a spring-mass system (Chap. 32) [3]. The ambient light sensor on the other hand has been proven to be a capable tool in studying an astronomical phenomenon [4] as well as in measuring speed and acceleration [5]. In this chapter we present an accurate, convenient, and engaging use of the smartphone magnetic field sensor to measure the acceleration due to gravity via measurement of the period of oscillations (simply called "period" in what follows) of a simple pendulum. Measurement of the gravitational acceleration via the simple pendulum is a standard elementary physics laboratory activity, but the employment of the magnetic field sensor of a smartphone device in measuring the period is quite new and the use of it is seen as fascinating among students. The setup and procedure are rather simple and can easily be replicated as a classroom demonstration or as a regular laboratory activity.

30.1 Theoretical Background

The simple pendulum is a classic example of a simple harmonic motion with its period being its most prominent parameter. Theoretically, the period of the pendulum T as a function of the length L and acceleration due to gravity g is expressed as [6]

U. Pili (✉) · R. Violanda · C. Ceniza
University of San Carlos, Cebu City, Cebu, Philippines
e-mail: ubpili@usc.edu.ph

© The Author(s), under exclusive license to Springer Nature Switzerland AG 2022
J. Kuhn, P. Vogt (eds.), *Smartphones as Mobile Minilabs in Physics*,
https://doi.org/10.1007/978-3-030-94044-7_30

$$T = 2\pi\sqrt{\frac{L}{g}}. \tag{30.1}$$

Equation (30.1) can be rewritten in the form

$$T^2 = \frac{4\pi^2}{g}L. \tag{30.2}$$

Equation (30.2) shows the linear relation between the square of the period and length with slope S:

$$S = \frac{4\pi^2}{g}. \tag{30.3}$$

Having gathered data points for the period and length, Eq. (30.2) can be plotted, T^2 against L. Substituting the slope of the resulting linear plot to Eq. (30.3), the acceleration due to gravity can be obtained.

30.2   The Experiment

The experimental setup, except for the computer and in part similar to a previous work [7], is depicted in Fig. 30.1. It is mainly composed of a smartphone with a built-in magnetic field sensor and is installed with an Android application called

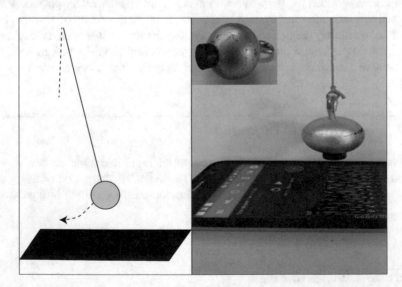

Fig. 30.1 The experimental setup (inset: small magnet glued at the bottom of the metallic ball). Although not used here, in a replicating experiment, it would be best to cover the smartphone with a hard transparent material in order to protect it should the pendulum accidentally fall off

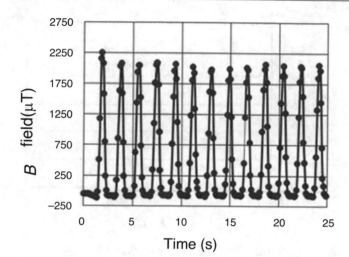

Fig. 30.2 Plot of B field against time. The varying values of the B field are due to the varying distance the magnet is located relative to the location of the B field sensor—in keeping with the distance dependence of the B field

Physics Toolbox Sensor Suites [8], a simple pendulum setup with a small magnet glued at the bottom of the pendulum bob, a tape measure, and a computer. Just like any other method, in the simple pendulum-based measurement of the acceleration due to gravity the crucial parameter needed to be measured directly and accurately as possible is the period. Using our setup in measuring the period, we launched the magnetic field sensor or magnetometer of the device via the application and then we allowed the pendulum, in the small angle regime, to oscillate a little above the screen of the device.

The pendulum bob was made to oscillate above the length side of the screen as seen in Fig. 30.1. In the process, a time series record of the data, as graphically depicted in Fig. 30.2, of the values of the magnetic field (simply called B field in what follows) was registered.

The data directly saved in MS Excel in the device were then transferred to the computer via email or Bluetooth, and from the computer-generated MS Excel graph such as the one shown in Fig. 30.2, the period of the pendulum was obtained. The device provides the x, y, and z-components of the B field (as well as their total magnitude) and any of these can be utilized depending on the prominence and clearness of the resulting time-based plots. The period from Fig. 30.2 is equal to the time interval between two successive peaks (or valleys) with the time coordinates obtained using the data cursor. We took the average period by taking into account at least 10 pairs of successive peak. Further, in order to increase the reliability of our results, we have considered eight (with length equal to zero automatically included) different lengths, allowing us to plot, also in MS Excel, the square of the period against length. The linear plot is depicted in Fig. 30.3.

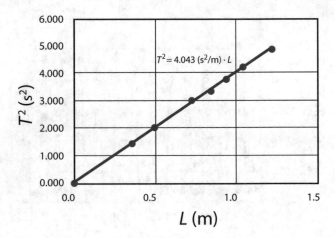

Fig. 30.3 Linear fit to the data points with slope 4.043 s²/m

Inserting the slope in Eq. (30.3) and using the MS Excel linear regression feature to obtain the error in the slope, then subsequently using error propagation for division [9], we found the value of the acceleration due to gravity to be 9.8 ± 0.1 m/s². This experimental value fits well within the local theoretical value [10] of 9.8 m/s².

30.3   Summary

The magnetic field sensor of the smartphone has emerged as a reliable timer for the measurement of the period of a simple pendulum and thus an accurate measurement of the acceleration due to gravity. This relatively exciting experiment, timer-wise, appears to be easy to replicate as a laboratory activity or, by not having to vary the length, can be used as a handy classroom demonstration setup for the verification of Eq. (30.1) via measurement of the acceleration due to gravity in just one run.

Acknowledgment We are indebted to J. R. Bahinting, Harold Bolanon, and Rommel Gomez, laboratory attendants of the Department of Physics, University of San Carlos, for providing us the magnetic simple pendulum setup. We also thank the University of San Carlos, owner of the simple pendulum setup, and the editors whose comments and suggestions improved the manuscript.

References

1. Countryman, C.: Familiarizing students with the basics of a smartphone's internal sensors. Phys. Teach. **52**, 557 (2014)
2. Vogt, P., Kuhn, J.: Analyzing free fall with a smartphone acceleration sensor. Phys. Teach. **50**, 182 (2012)
3. Kuhn, J., Vogt, P.: Analyzing spring pendulum phenomena with a smartphone acceleration sensor. Phys. Teach. **50**, 504 (2012)

4. Barrera-Garrido, A.: Analyzing planetary transits with a smartphone. Phys. Teach. **53**, 179 (2015)
5. Kapucu, S.: Finding the acceleration and speed of a lightemitting object on an inclined plane with a smartphone light sensor. Phys. Educ. **52**, 055003 (2017)
6. Young, H.D., Freedman, R.A., Ford, A.L.: Sears and Zemansky's University Physics, 13th edn, p. 454. Addison-Wesley (2012)
7. Sinacore, J., Takai, H.: Measuring *g* using a magnetic pendulum and telephone pick up. Phys. Teach. **48**, 448–449 (2010)
8. We have downloaded the Android application Physics Toolbox Suites for free from Google Play.
9. Deacon, C.: Error analysis in the introductory physics laboratory. Phys. Teach. **30**, 368 (1992)
10. We have obtained our local theoretical value of the acceleration due to gravity by inserting our altitude and longitude into a formula that appears at this website: https://www.npl.co.uk/resources/q-a

Determination of Gravity Acceleration with Smartphone Ambient Light Sensor

31

Jhon Alfredo Silva-Alé

Among the serious experimental uses given to smartphones in physics, the use of integrated sensors, video analysis, and mixed uses as simultaneous material and instruments stand out for their frequency. Regarding the use of integrated sensors, there are many proposals that use the accelerometer (e.g. Chaps. 8 and 32) [1–3], the magnetometer (e.g. Chaps. 30 and 34) [4, 5], or other sensors (e.g., camera (Chap. 27) [6] or microphone [e.g. Chap. 12) [7–9]] to find experimental values of physical constants, such as acceleration of gravity (g), thereby trying to contribute to the studies of statics, kinematics, and dynamics. However, there are few studies that use the ambient light sensor for this kind of purpose.

This chapter proposes to establish an experimental value of the acceleration constant of gravity (g) using the ambient light sensor integrated in the smartphone, with the help of a simple pendulum with an experimental configuration, in order to analyze the possibility of using it as a measurement instrument.

31.1 Theoretical Background

When the spherical mass of the pendulum oscillates in a small arc of circumference through a slightly viscous medium, such as air, it describes a simple and weakly damped harmonic movement. In this type of analysis, it is usual to find experimental proposals that neglect the friction of the sphere with air, to focus on the relationship between the oscillation period (T), the length of the rope (L), and the gravitational acceleration (g). This relationship is expressed as:

J. A. Silva-Alé (✉)
USACH, UCH, Santiago, Chile
e-mail: jhon.silva@usach.cl

J. Kuhn, P. Vogt (eds.), *Smartphones as Mobile Minilabs in Physics*,
https://doi.org/10.1007/978-3-030-94044-7_31

$$T = 2\pi\sqrt{\frac{L}{g}}. \tag{31.1}$$

Using Eq. (31.1), it is possible to calculate a theoretical value of the period of oscillation (T_{th}) by setting a value for L and considering the theoretical magnitude of the acceleration due to gravity: $g = 9.806$ m/s^2.

To collect data from the experimental period (T_{exp}), the ambient light sensor is used. To determine an experimental value associated with the magnitude of the acceleration due to gravity (g_{exp}), Eq. (31.1) has been rewritten as follows:

$$g_{exp} = \frac{4\pi^2 L}{T_{exp}^2}. \tag{31.2}$$

The measurements of L and T_{exp} are used in it.

As the amount of data collected is slightly more than 10 units, the estimating measurement uncertainty (Δx) has been used based on the sensitivity and the mean range, which is presented in Eq. (31.3).

$$\Delta x = \frac{\text{Sensitivity}}{2} + \frac{\text{Lim}_{sup} - \text{Lim}_{inf}}{2}. \tag{31.3}$$

31.2 The Proposal

Figure 31.1 presents the experimental setup used to carry out the experiment. The assembly consists of a constant light source, a simple pendulum system, and a smartphone with an integrated ambient light sensor, to which the mobile application Physics ToolBox Suite has been installed [10].

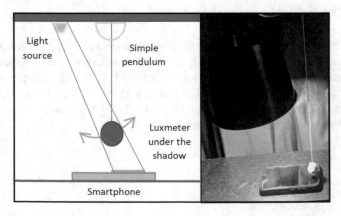

Fig. 31.1 Experimental setup. The image on the left is an illustration of the components involved in the assembly, while the image on the right is a real photo

Using a millimeter ruler, the length of the rope $L = 36.5$ cm, from the surface of the sphere to the top knot, has been measured. Then the Physics Toolbox application is configured to function as a light meter. The sensor is placed under the shadow cast by the sphere. Then the record button is pressed to start capturing the differences in light intensity.

To produce these variations the sphere is displaced approximately $10°$ from the equilibrium position.

31.3 Processing and Analysis of Results

The data of the variation of the light intensity as a function of time have been recorded by letting the pendulum swing for 10 s. These data were immediately exported from the smartphone to the computer, through a CSV file in order to be organized and graphed in an Excel file, as shown in the graph in Fig. 31.2.

Subsequently, the T_{exp} of the 13 oscillations performed during the 10-second follow-up were determined, as presented in Table 31.1. For this, the time difference between consecutive peaks and/or valleys was applied.

After obtaining each of the T_{exp}, the T_{th} was calculated using Eq. (31.1):

$$T_{th} = 2\pi\sqrt{\frac{0.365\,\text{m}}{9.806\,\frac{\text{m}}{\text{s}^2}}} = 1.212\,\text{s}.$$

Using Eq. (31.2), the error percentage was determined by comparing each of T_{exp} with T_{th}. Finally, using the value T_{exp}, and the measured value of L, g_{exp} has been obtained:

$$g_{exp} = \frac{4\pi(0.365\,\text{m})}{(1.217\,\text{s})^2} = 9.729 \pm 0.01\,\frac{\text{m}}{\text{s}^2}.$$

NOTE: 0.01 is due to the error propagation in measurement uncertainty ΔT_{exp}^2.

Fig. 31.2 Graph of perceived light intensity (lx) vs. time (s). The variation of the light intensity value is due to the shadow projections of the sphere

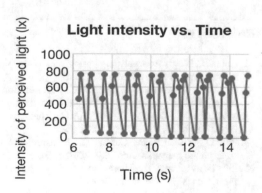

Table 31.1 Table summarizing the experimental periods obtained from the 13 oscillations and the percentage error of the comparison with the theoretical period

$T_{exp} \pm 0.0085$ (s)	ε %
1.212	0.0
1.216	0.4
1.217	0.4
1.217	0.4
1.220	0.7
1.216	0.4
1.217	0.4
1.218	0.5
1.215	0.3
1.214	0.2
1.218	0.5
1.219	0.6
1.216	0.4
$\overline{T}_{exp} = 1.217$	$\overline{\varepsilon} = 0.4$

The Standard Error (SE) value of g_{exp} is 0.009 (approximately 0.01). The above shows low variability of the g_{exp} measurements.

When comparing the value of g_{exp} with the local theoretical value g_{th}, the percentage error is less than 0.785%. Despite the low percentage error, g_{th} is not in the error range of g_{exp}. Therefore, the experimental value obtained for the calculation of the acceleration of gravity fits well as an approximate value to the theoretical value.

31.4   Conclusions

Thanks to this experience, it was possible to determine that the experimental value of the gravity acceleration (g_{exp}) obtained by using the smartphone's ambient light intensity sensor, during the simple pendulum tracking, fits well with the theoretical value.

References

1. Kuhn, J., Vogt, P.: Analyzing spring pendulum phenomena with a smartphone acceleration sensor. Phys. Teach. **50**, 504 (2012 Oct)
2. Monteiro, M., Cabeza, C., Martí, A.C.: Acceleration measurements using smartphone sensors: dealing with the equivalence principle. Rev. Bras. Ens. Fís. **37**, 1303 (2015 Aug)
3. Monteiro, M., Cabeza, C., Martí, A.C.: The Atwood machine revisited using smartphones. Phys. Teach. **53**, 373–374 (2015 Sept)
4. Pili, U., Violanda, R., Ceniza, C.: Measurement of g using a magnetic pendulum and a smartphone magnetometer. Phys. Teach. **56**, 258–259 (2018 March)
5. Pili, U., Violanda, R.: Measuring a spring constant with a smartphone magnetic field sensor. Phys. Teach. **57**, 198 (2019 Feb)

6. Becker, S., Klein, P., Kuhn, J.: Video analysis on tablet computers to investigate effects of air resistance. Phys. Teach. **54**, 440–441 (2016 Oct)
7. Kuhn, J., Vogt, P.: Smartphone & Co. in Physics Education: Effects of Learning with New Media Experimental Tools in Acoustics. In: Schnotz, W., Kauertz, A., Ludwig, H., Müller, A., Pretsch, J. (eds.) Multidisciplinary Research on Teaching and Learning, pp. 253–269. Palgrave Macmillan, Basingstoke (2015)
8. Kuhn, J., Vogt, P.: Smartphones as experimental tools: different methods to determine the gravitational acceleration in classroom physics by using everyday devices. Eur. J. Phys. Educ. **4**(1), 16–27 (2013)
9. Schwarz, O., Vogt, P., Kuhn, J.: Acoustic measurements of bouncing balls and the determination of gravitational acceleration. Phys. Teach. **51**, 312–313 (2013 May)
10. https://ogy.de/physicstoolboxsensorsuite

Analyzing Spring Pendulum Phenomena with a Smartphone Acceleration Sensor

32

Jochen Kuhn and Patrik Vogt

This chapter describes two further pendulum experiments using the acceleration sensor of a smartphone in this book [for earlier contributions concerning this topic, including the description of the operation and use of the acceleration sensor, see Refs. [1] (Chap. 6) and [2] (Chap. 29)]. In this chapter we focus on analyzing spring pendulum phenomena. Therefore two spring pendulum experiments will be described in which a smartphone is used as a pendulum body and SPARKvue [3] software is used in conjunction with an iPhone or an iPod touch, or the Accelogger [4] app for an Android device [1, 2]. As described in Ref. [1], the values measured by the smartphone are subsequently exported to a spreadsheet application (e.g., MS Excel) for analysis.

32.1 The Spring Pendulum

A smartphone can also be used as a swinging mass to record the oscillation of a spring pendulum. A measurement example that was recorded with an iPhone with a mass $m = 0.152$ kg is shown in Fig. 32.1. It results in an almost perfect sinusoidal process, which offers similar learning opportunities as the results of the string pendulum:

(a) Why do the accelerations vary around g and at what points of the swing are the maxima and minima located?

J. Kuhn (✉)
Ludwig-Maximilians-Universität München (LMU Munich), Faculty of Physics, Chair of Physics Education, Munich, Germany
e-mail: jochen.kuhn@lmu.de

P. Vogt
Institute of Teacher Training (ILF) Mainz, Mainz, Germany
e-mail: vogt@ilf.bildung-rp.de

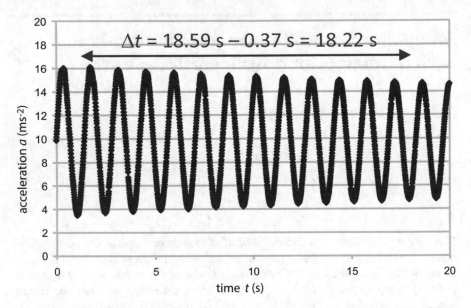

Fig. 32.1 Dynamic determination of spring constants: chronological acceleration process for a spring pendulum (Presentation of measured values after the export of data from the smartphone into MS Excel)

(b) What is the duration of a period of an oscillation and how high is the spring constant of the spring used?

(c) How strongly is the oscillation dampened?

Examples related to items (a) and (b) will be discussed, whereby the dynamically determined spring constant will be compared with the result of a static measurement. As regards (a), as the pendulum slows down (motion away from the rest point) and as it speeds up (motion toward the rest point), inertia forces evolve, which are recorded in addition to the weight. The maximum accelerations occur at the top turning point of the swing, the minimum accelerations with maximum extension of the springs.

As regards b), on the basis of the data underlying Fig. 32.1, 13 complete oscillations occur in a time period $\Delta t = 18.22$ s.

With

$$T = \frac{\Delta t}{n} = 2\pi\sqrt{\frac{m}{D}},$$ (32.1)

(n = number of oscillations in time interval Δt), the spring constant D of the used spring can be determined by applying the formula

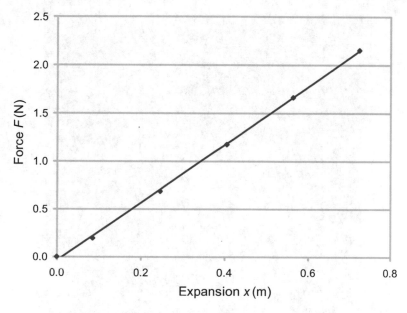

Fig. 32.2 Static determination of spring constants: the weight F impacting the spring as a function of its elongation x

$$D = \frac{4\pi^2 m}{\left(\frac{\Delta t}{n}\right)^2} = (3.055 \pm 0.003)\,\frac{\text{N}}{\text{m}}.$$

This value matches well with the result of a static measurement, in which the displacement x of the spring was examined in relation to the load F applied to it (forces resulting from the attached mass). For an elastic spring, Hooke's law of elasticity is applied

$$F = D \cdot x, \tag{32.2}$$

i.e., the F-x plot results in a straight line, with the slope corresponding to the spring constants (Fig. 32.2). A linear regression confirms the linear relationship with an adjusted coefficient of determination to close to one and determines the spring constant to be $3.02\ \text{Nm}^{-1}$.

32.2 Coupled Pendulum

If two spring pendula are coupled together by an attached mass (Fig. 32.3) and one of the two systems is set into motion, the oscillation performed by that pendulum soon also transfers over to the other pendulum. Provided that the springs used possess the same stiffness—i.e., the same spring constant—the first pendulum remains briefly motionless after some time; its oscillation energy has been completely transferred

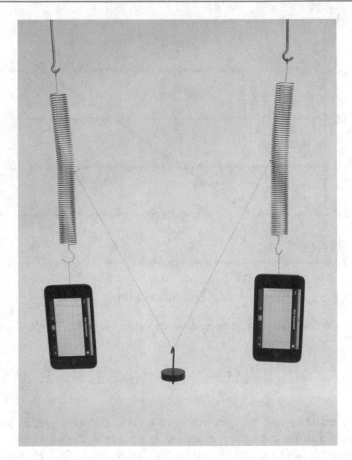

Fig. 32.3 Experiment setup for a coupled pendulum

over to the second pendulum. From then on, the energy shifts between the two pendula, which is why each of the pendula makes a typical swing motion.

The oscillations of coupled pendula of this kind can easily be recorded with the help of two smartphones, which once again serve as mass for the pendulum (Fig. 32.3). A measurement example for a coupled mass of 100 g is shown in Fig. 32.4. When the acceleration values, and therefore the amplitudes, of one pendulum reach a maximum, the accelerations of the other pendulum correspond approximately to the acceleration of gravity, which equals a state of rest. The energy transfer happens even faster if the two pendula are more strongly coupled together. By varying the coupled mass, this aspect can be examined in a subsequent experiment.

There is a complication when analyzing the data because the measurements of both smartphones cannot be started completely synchronously. In this case, the following method has proved successful: The person conducting the experiment starts the measurements on both smartphones separately and waits for the systems to

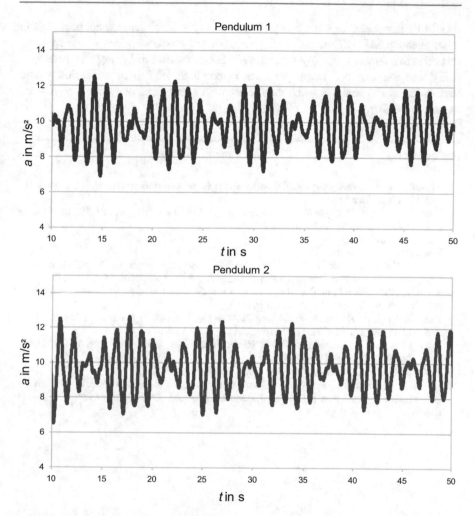

Fig. 32.4 Acceleration process of both coupled spring pendula (Presentation of measured values after the export of data from the smartphone into MS Excel) [5]

reach a state of rest. Then he or she lightly taps the stand, which later clearly appears as a peak in both series of measurements. When analyzing the data, the time axes can then be moved so that the acceleration maxima resulting from the tap are positioned exactly on top of each other, corresponding to synchronization.

The following aspect should be emphasized before starting: If the lesson objective is to determine the period of an oscillation in an experiment, it is advisable to use a conventional stopwatch rather than a smartphone. The use of a cell phone as a pendulum body, however, generates a much greater quantity of information and learning opportunities than the duration of a period alone and can greatly enhance instruction. So, studying this phenomenon using smartphones' acceleration sensor should be integrated in a more sophisticated instructional setting, connected with other phenomena [6–8] and addressing the students' misconceptions which could be related to this concept [9]. Anyway, current research shows that using this method to study free fall and oscillation phenomena could at least increase curiosity and motivation of the students when they learn with smartphones or tablets as experimental tools [10].

References

1. Vogt, P., Kuhn, J.: Analyzing free fall with a smartphone acceleration sensor. *Phys. Teach.* **50**, 182–183 (2012 March)
2. Vogt, P., Kuhn, J.: Analyzing simple pendulum phenomena with a smartphone acceleration sensor. *Phys. Teach.* **50**, 439–440 (2012 Oct)
3. https://ogy.de/sparkvue
4. https://ogy.de/accelogger
5. Vogt, P., Kuhn, J., Gareis, S.: Beschleunigungssensoren von Smartphones: Möglichkeiten und Beispielexperimente zum Einsatz im Physikunterricht. *PdN-PhiS.* **7**(60), 15–23 (2011)
6. Kuhn, J.: Relevant information about using a mobile phone acceleration sensor in physics experiments. *Am. J. Phys.* **82**(2), 94 (2014)
7. Kuhn, J., Vogt, P., Müller, A.: Analyzing Elevator Oscillation with the Smartphone Acceleration Sensors. *Phys. Teach.* **52**(1), 55–56 (2014)
8. Vogt, P., Kuhn, J.: Analyzing collision processes with the smartphone acceleration sensor. *Phys. Teach.* **52**(2), 118–119 (2014)
9. Hall, J.: More smartphone acceleration. *Phys. Teach.* **51**(1), 6 (2013)
10. Hochberg, K., Kuhn, J., Müller, A.: Using smartphones as experimental tools – effects on interest, curiosity and learning in physics education. *J. Sci. Educ. Technol.* **27**(5), 385–403 (2018)

Using the Smartphone as Oscillation Balance

33

A. Kaps and F. Stallmach

The dynamics of a spring pendulum is an important topic in introductory experimental physics courses at universities and advanced science courses in secondary school education. Different types of pendulum setups with smartphones were proposed to investigate, e.g., the gravitational acceleration [1, 2], to measure spring constants, or to illustrate and verify the principles of periodic motions and their characterization via oscillation periods [3–8]. In most of these experiments the smartphone with its internal MEMS sensors or the magnetometer is used as the measurement device.

In this chapter we present an experiment where students may apply the concept of the harmonic oscillations to determine the unknown mass of an object with a so-called oscillation balance, in which a dynamic method of measuring masses is realized. In our experimental setup the smartphone represents the measurement device and the pendulum bob with a known mass. We use the app *phyphox* (RWTH Aachen, Germany) to record and transfer the data of the internal MEMS accelerometer [9, 10].

The advantage of this dynamic measurement principle is that unknown masses of objects may be determined independent of the local gravitational force. However, if realized using a spring pendulum, all parts performing the periodic motion, including the smartphone, its fastening materials to the spring, and a fraction of the spring mass itself [11–13], will contribute to the mass measured and need to be considered in data analysis.

A. Kaps (✉) · F. Stallmach
Department Didactics of Physics, Faculty for Physics and Earth Sciences, University Leipzig, Leipzig, Germany
e-mail: andreas.kaps@uni-leipzig.de; stallmac@physik.uni-leipzig.de

33.1   Theoretical Background

The spring pendulum is an example of a system undergoing simple harmonic oscillations. Its oscillation period T is a function of the mass m and of the spring constant D. It is

$$T = 2\pi \cdot \sqrt{\frac{m}{D}}. \tag{33.1}$$

In our oscillation balance m is the sum of the masses of all objects attached to the spring. It also includes the effective mass $f \cdot m_f$ of the spring itself [11–13], where $f = 1/2$ for the case of a soft spring, which we used in our setup. The oscillation period is measured twice: First T_0 with a known mass m_0, consisting of the smartphone including its fastening materials. Second T_1 with an object of unknown mass m_e attached additionally to the same spring (Fig. 33.1). Thus, the ratio of the square of the oscillation periods of T_1^2/T_0^2 depends only on the ratio of total masses performing the oscillation.

$$\frac{T_1^2}{T_0^2} = \frac{m_0 + f \cdot m_f + m_e}{m_0 + f \cdot m_f}. \tag{33.2}$$

Equation (33.2), in which the unknown spring constant D is eliminated, is resolved for the mass m_e:

$$m_e = (m_0 + f \cdot m_f) \cdot \left(\frac{T_1^2}{T_0^2} - 1\right), \tag{33.3}$$

which now can be measured via the two oscillation periods.

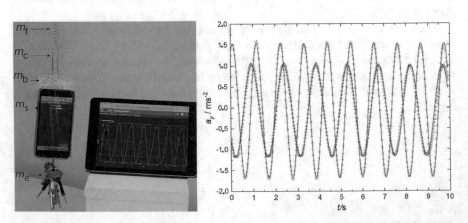

Fig. 33.1   Top: Experimental setup for the oscillation balance. Bottom: Recorded acceleration $a_y(t)$ in dependence on time t. The blue and red data points represent the oscillations with and without additional mass m_e, respectively. The lines describe the fits via Eq. (33.4)

33.2   The Experiment

The experiment is performed by attaching the smartphone with mass m_s or the smartphone and the object with unknown mass $m_s + m_e$ via a reclosable bag with a paper clip to the spring (Fig. 33.1, top and masses in Table 33.1). In order to fulfill Hooke's law, only small oscillation amplitudes are excited. Several stable periods of the acceleration of the smartphone are recorded via the app phyphox. To obtain the oscillation periods with high accuracy, the data are imported to a scientific software and analyzed via an undamped sine fit function

$$a_y(t) = A \cdot \sin\left(\frac{2\pi}{T} \cdot t + \varphi_0\right),\qquad (33.4)$$

where A and φ_0 denote the amplitude and the phase shift of the acceleration, respectively (Fig. 33.1, bottom).

The validity of the measurement principle was checked by varying m_e from 0 to 300 g. As expected from Eq. (33.2) and depicted in Fig. 33.2, we obtained a linear variation of the square of the ratio of the oscillation periods in dependence on the variable mass m_e, which demonstrates the working principle of the oscillation balance. Generally, our students perform this experiment with objects of unknown masses of their choice, for example with their own keychain attached to the spring-smartphone setup described (Fig. 33.1, top). Table 33.1 provides the resulting oscillation period T_1. Via Eq. (33.3) the mass of the keychain is calculated to be $m_e = (236 \pm 30)$ g. Within the computed uncertainties, this result agrees with the reference value of $m_{e,ref} = (256.10 \pm 0.02)$ g determined with a conventional digital balance.

In conclusion, these results indicate that the smartphone and its internal acceleration sensor can be utilized to realize a dynamic measurement of masses. Compared to a commercial digital balance, the proposed smart(phone) oscillation balance is less accurate since it depends strongly on the square ratio of two oscillation periods and their uncertainties (Fig. 33.2). In addition current research shows that especially the topic of oscillations is suitable to increase curiosity and motivation [14] as well as conceptual understanding [15] of the students when they learn with smartphones or tablets as experimental tools [16].

Table 33.1 Masses for the oscillation balance experiment measured with a digital balance and oscillation periods T_0 and T_1 for the experiment with the keychain	

Mass of the smartphone m_s	(211.70 ± 0.02) g
Mass of the reclosable bag m_b	(1.00 ± 0.02) g
Mass of the paper clip m_c	(3.00 ± 0.02) g
Total mass m_0	(215.7 ± 0.1) g
Mass of the spring m_f	(3.85 ± 0.02) g
Oscillation period T_0	(1.01 ± 0.2) s
Oscillation period T_1	(1.46 ± 0.02) s

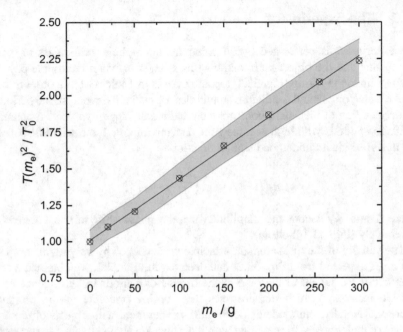

Fig. 33.2 Linear fit ($R^2 > 0.99$) of the measured oscillation period data $T^2(m_e)/T_0^2$ for $0 \leq m_e \leq 300$ g. The blue area represents the computed uncertainty for the T_1^2/T_0^2 ratio

Acknowledgments The authors are grateful for financial support received via the STIL project of the University of Leipzig (grant number 01PL16088, BMBF Germany).

References

1. Pili, U.: Measurement of g using a magnetic pendulum and a smartphone magnetometer. Phys. Teach. **56**, 258–259 (2018 April)
2. Pili, U., Violanda, R.: Measuring a spring constant with a smartphone magnetic field sensor. Phys. Teach. **57**, 198–199 (2019 March)
3. Kuhn, J., Vogt, P.: Analyzing spring pendulum with a smartphone acceleration sensor. Phys. Teach. **50**, 439–440 (2012 Oct)
4. Kuhn, J., Vogt, P., Mueller, A.: Analyzing elevator oscillation with the smartphone acceleration sensors. Phys. Teach. **52**, 55–56 (2014 Jan)
5. Palacio, J.C., Velazquez-Abad, L., Monsorio, G.J.A.: Using a mobile phone acceleration sensor in physics experiments on free and damped harmonic oscillations. Am. J. Phys. **81**, 472–475 (June 2013)
6. Weiler, D., Bewersdor, A.: Superposition of oscillation on the Metapendulum: visualization of energy conservation with the smartphone. Phys. Teach. **57**, 646–647 (2019 Dec)
7. Kaps, A., Splith, T., Stallmach, F.: Shear modulus determination using the smartphone in a torsion pendulum. Phys. Teach. **59**, 268 (2021). https://doi.org/10.1119/10.0004154
8. Kaps, A., Stallmach, F.: Smart physics with an oscillating beverage can. Phys. Educ. **56**(4), 045010 (2021).

9. Staacks, S., Hütz, S., Heinke, H., Stampfer, C.: Advanced tools for smartphone-based experiments: phyphox. Phys. Educ. **53**, 045009 (2018 July)
10. https://phyphox.org/de/home-de/
11. Ruby, L.: Equivalent mass of a coil spring. Phys. Teach. **38**, 140–142 (2000 March)
12. Green, N., Gill, T., Everly, S.: Finding the effective mass and spring constant of a force probe from simple harmonic motion. Phys. Teach. **54**, 138–142 (2016 March)
13. Rodriguez, E., Gesnouin, G.: Effective mass of an oscillating spring. Phys. Teach. **45**, 100–104 (2007 Feb)
14. Hochberg, K., Kuhn, J., Müller, A.: Using smartphones as experimental tools – Effects on interest, curiosity and learning in physics education. J. Sci. Educ. Technol. **27**(5), 385–403 (2018)
15. Hochberg, K., Becker, S., Louis, M., Klein, P., Kuhn, J.: Using smartphones as experimental tools – a follow-up: cognitive effects by video analysis and reduction of cognitive load by multiple representations. J. Sci. Educ. Technol. **29**(2), 303–317 (2020)
16. Kaps, A., Splith, T., Stallmach, F.: Implementation of smartphone-based experimental exercises for physics courses at universities. Phys. Educ. **56**(3), 0035004 (2021). https://doi.org/10.1088/1361-6552/abdee2

Measuring a Spring Constant with a Smartphone Magnetic Field Sensor

34

Unofre Pili and Renante Violanda

In introductory physics laboratories, spring constants are traditionally measured using the static method. The dynamic method, via vertical spring-mass oscillator, that uses a stopwatch in order to measure the period of oscillations is also commonly employed. However, this time-measuring technique is prone to human errors and in this chapter we present a similar setup, except for the motion timer being the B-field sensor of a smartphone, in order to measure the period of oscillations and thus the spring constant. The smartphone device as an introductory physics experimental tool is quite well established [1]. For instance, the smartphone-based acceleration sensor has been employed in a rather quick measurement of the acceleration due to gravity (Chap. 6) [2] and in analyzing simple pendulum phenomena (Chap. 29) [3]. In addition, the magnetic field sensor of the smartphone device has been effectively used as well in measuring average angular velocity (Chap. 14) [4] and the acceleration due to gravity (Chap. 30) [5].

34.1 Theoretical Background

The oscillations of a spring-mass oscillator is one classic example of simple harmonic motion in which the period of oscillations T is expressed as [6]

$$T = 2\pi\sqrt{\frac{m + m_{\text{eff(s)}}}{k}},$$

(34.1)

where m is the suspended mass, k the spring constant, and $m_{\text{eff(s)}}$ is the effective mass of the spring and is equal to one-third of the mass of the spring for the case in which

U. Pili (✉) · R. Violanda
University of San Carlos, Cebu City, Cebu, Philippines
e-mail: ubpili@usc.edu.ph

© The Author(s), under exclusive license to Springer Nature Switzerland AG 2022
J. Kuhn, P. Vogt (eds.), *Smartphones as Mobile Minilabs in Physics*,
https://doi.org/10.1007/978-3-030-94044-7_34

the ratio between the suspended mass and the mass of the spring is substantially greater than one [6]. Equation (34.1) can be written as

$$T^2 = \frac{4\pi^2}{k} m_{\mathrm{T}}, \tag{34.2}$$

where $m_{\mathrm{T}} = m + m_{\mathrm{eff(s)}}$ is the total suspended mass. Equation (34.2) presents a linear relationship between the square of the period T_2 and the total suspended mass m_{T} with slope S:

$$S = \frac{4\pi^2}{k}. \tag{34.3}$$

With the availability of data points for T_2 and m_{T}, and subsequently plotting T_2 against m_{T}, the spring constant can be determined from the slope—of the resulting linear plot—by substituting it in Eq. (34.3).

34.2   Experiment

The experimental setup (except for the computer) is shown in Fig. 34.1. It is made up of a vertical spring-mass oscillator acquired from PHYWE (with extra 50-g mass slugs necessary for varying the total suspended mass of the system), a computer with

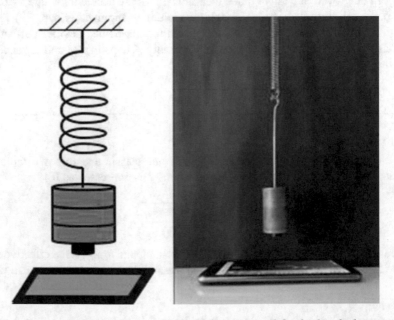

Fig. 34.1 Experimental setup. The small cylindrical magnet, at left, glued at the bottom of the mass hanger is emphasized. Protecting the phone with a hard transparent material is recommended, in case the suspended mass accidentally falls off

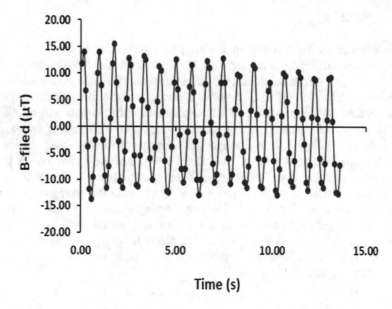

Fig. 34.2 Time series plot of the B-field corresponding to one of the suspended masses. It does reflect the periodic motion of the suspended mass; the variation of the magnetic field is simply due to its dependence on distance

MS Excel, and a smartphone installed with an Android application called Physics Toolbox Sensor Suites [7]. In addition, a small permanent magnet is attached at the bottom of the mass hanger. The primary parameter with which this setup is intended is to measure the period of oscillations of the oscillator.

Performing the experiment, we launched the B-field sensor of the smartphone via the Android application and subsequently allowed the spring-mass oscillator to oscillate a couple of centimeters above the smartphone screen, generating time series plots (x-, y-, and z-components) of the B-field. We then imported the MS Excel-compatible CSV file into the computer from which we regenerated the plots and did further and necessary data analysis. To finally obtain the period of oscillations, we chose the z-component of the B-field because it emerged to be the most prominent as far as our setup was involved. We have taken into account six different values of the suspended mass, including the mass of the magnet and hanger, in order to gather, as well as increase, the accuracy of the results, data points for T^2 and m_T. Figure 34.2 presents one of the six time-based B-field plots, corresponding to one of the suspended masses that we regenerated in MS Excel.

34.3   Results

The time separation between two successive peaks (or valleys) in the B-field against time plot shown in Fig. 34.2 is the period of oscillations. The value of the period is equal to the time difference obtained after subtracting the time coordinate of the preceding peak (or valley) from the time coordinate of the immediate succeeding peak (or valley). We have obtained the time coordinate of each peak with the help of the data cursor. We took the average period by considering 10 pairs of two successive peaks (or valleys). Each average period corresponding to each of the six different suspended masses already obtained, we plotted the square of the period against the suspended mass. The resulting linear plot is shown in Fig. 34.3.

Inserting the value of the slope in Eq. (34.3), we found the average value of the spring constant to be 29.5 N·m^{-1}. Then performing linear regression on the data points, we obtained the standard error in the slope and, subsequently applying error propagation for division [8] on Eq. (34.3), we found the spring constant to be within 29.5 ± 0.1 N·m^{-1}. The indicated accepted commercial value of the spring constant of the spring we used is 30.0 N·m^{-1}.

34.4   Conclusions

We have demonstrated that the magnetic field sensor of a smartphone is a reliable motion timer. It appears to be novel, in the experiment presented at least, and accurate. Because smartphone devices are popularly known for their everyday uses and not for their applications in scientific experiments, it is seen that students will likely find fun and fascination performing the experiment themselves.

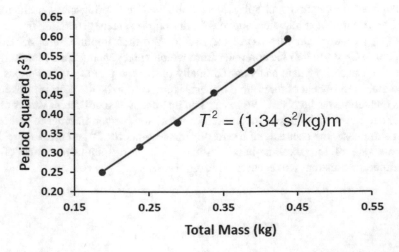

Fig. 34.3 Plot of the square of the period against total suspended mass. The red line is the best fit line to the data points with slope equal to 1.34 s^2·kg^{-1}

Furthermore, other experiments involving motion can be explored using the magnetic field sensor of smartphones for time interval measurements.

Acknowledgments   We are thankful to University of San Carlos for its continued support, in terms of research and publication, to its faculty. The setup is owned by the Department of Physics and was provided for us by Mr. J. R. Bahinting. We also thank the editor and the anonymous reviewer for their valuable suggestions.

References

1. Countryman, C.: Familiarizing students with the basics of a smartphone's internal sensors. Phys. Teach. **52**, 557 (2012 Dec)
2. Vogt, P., Kuhn, J.: Analyzing free fall with a smartphone acceleration sensor. Phys. Teach. **50**, 182 (2012 March)
3. Vogt, P., Kuhn, J.: Analyzing simple pendulum phenomena with a smartphone acceleration sensor. Phys. Teach. **50**, 439 (2012 Oct)
4. Pili, U., Violanda, R.: Measuring average angular velocity with a smartphone magnetic field sensor. Phys. Teach. **56**, 114–115 (2018 Feb)
5. Pili, U., Violanda, R., Ceniza, C.: Measurement of g using a magnetic pendulum and smartphone magnetometer. Phys. Teach. **56**, 258–259 (2018 April)
6. Rodriguez, E.E., Gesnouin, G.A.: Effective mass of an oscillating spring. Phys. Teach. **45**, 100 (2007 Jan)
7. We have downloaded for free the Android application Physics Toolbox Sensor Suites from Google Play
8. Deacon, C.: Error analysis in the introductory physics laboratory. Phys. Teach. **30**, 368 (1992 Sept)

Analyzing Elevator Oscillation with the Smartphone Acceleration Sensors

35

Jochen Kuhn, Patrik Vogt, and Andreas Müller

It has often been reported in this book that smartphones are very suitable tools for exploring the physical properties of everyday phenomena. A very good example of this is an elevator ride. In addition to the acceleration processes, oscillations of the cabin are interesting. The present work responds to the second aspect.

35.1 Theoretical Background and Execution of the Experiment

When riding in a cable-driven elevator, you might notice that a sudden jolt, perhaps from someone jumping vertically in the elevator, causes the elevator to start swaying or oscillating, with the period of oscillation decreasing at higher levels.

The reason for this is that, in simplified terms, a cable-driven elevator can be seen as a spring pendulum. Several previous contributions have described how a spring pendulum can be explored using smartphones (Chap. 32) [1–3]. The elevator cable corresponds to the spring of a spring pendulum with the spring constant k and the elevator car corresponds to the pendulum body M. The duration of periods T is therefore calculated with the equation

J. Kuhn (✉)
Ludwig-Maximilians-Universität München (LMU Munich), Faculty of Physics, Chair of Physics Education, Munich, Germany
e-mail: jochen.kuhn@lmu.de

P. Vogt
Institute of Teacher Training (ILF) Mainz, Mainz, Germany
e-mail: vogt@ilf.bildung-rp.de

A. Müller
Université de Genève, Fac. des Sciences/Sect. Physique, Institut Universitaire de Formation des Enseignants (IUFE), Pavillon d'Uni Mail (IUFE), Genève, Switzerland
e-mail: Andreas.Mueller@unige.ch

© The Author(s), under exclusive license to Springer Nature Switzerland AG 2022
J. Kuhn, P. Vogt (eds.), *Smartphones as Mobile Minilabs in Physics*,
https://doi.org/10.1007/978-3-030-94044-7_35

Fig. 35.1 Smartphone at the bottom of the elevator (here, iPhone with the SPARKvue app)

$$T = 2\pi\sqrt{\frac{M}{k}}.$$

The spring constant itself is, in turn, inversely proportional to the length L and directly proportional to the modulus of elasticity E and the cross section A of the elevator cable. Thus, the following applies

$$k = \frac{E \cdot A}{L}.$$

If this relationship is integrated into the first equation, the result is

$$T = 2\pi\sqrt{\frac{M \cdot L}{E \cdot A}} \rightarrow T^2 \propto L.$$

In order to explore the validity of this relationship, a smartphone or a tablet PC is placed on the floor of the elevator (Fig. 35.1). Next, while the elevator is stationary, the vertically acting component of acceleration (component in the z-direction, perpendicular to the display) is selected using an appropriate app (iOS: e.g. SPARKvue [4]; Android: e.g. Accelogger [5]) and the measurement is started. When a person jumps one time in the elevator, it causes it to start oscillating, which is then recorded using the app.

After importing the data into a spreadsheet, the period of oscillation is then determined from the oscillation curve of the app and the procedure is repeated on every story of the high-rise building (Fig. 35.2). In order to see the behavior

Fig. 35.2 Eleven-story high-rise building with a cable-driven elevator

predicted above, the high-rise building needs to have a sufficiently high number of stories (at least 10).

35.2   Experiment Analysis

Table 35.1 contains the results of measurements made as an example in the high-rise building in Fig. 35.2. For the experiment, the cable length L was estimated by adding together the heights of the individual stories and the estimated remaining height between the top floor and the building's roof. In the example described here, the story height h was measured as 3 m and the remaining length of the cable L_0 (length of cable at the top floor) was estimated to be 0.5 m.

If the square of the duration of the periods is entered against the estimated cable length in a system of coordinates, it is possible to explore how well the model above works (we found a coefficient of determination $R^2 > 0.94$; see Fig. 35.3). This measurement process was repeated in several different elevators; the experiment consistently produced similar results.

Figure 35.3 also clearly shows that the oscillation does not only depend on the pendulum length L, because if this was the case it would result in a line from the origin. Indeed, the ordinate intercept deviates significantly from zero in Fig. 35.3. A

Table 35.1 Example of experiment

Number of stories	Number of stories, counted starting from the top n	T in s	T^2 in s^2	$L = n \cdot h + L_0$
11	0	0.15	0.02	0.50
10	1	0.16	0.03	3.46
9	2	0.20	0.04	6.42
8	3	0.22	0.05	9.38
7	4	0.22	0.05	12.34
6	5	0.24	0.06	15.30
5	6	0.26	0.07	18.26
4	7	0.26	0.07	21.22
3	8	0.27	0.07	24.18
2	9	0.27	0.07	27.14
1	10	0.28	0.08	30.10

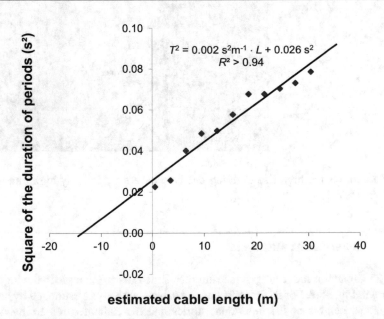

$$T^2 = 0.002 \ s^2m^{-1} \cdot L + 0.026 \ s^2$$
$$R^2 > 0.94$$

Fig. 35.3 Graphical depiction of the measured values

possible reason for this is, on the one hand, that an elevator system is generally equipped with shock absorbers. The additional elasticity of the unrolled cable results in an extension of the duration of oscillation. An additional influencing factor could be that the cable is directed around a fixed pulley and that the estimated length of the cable therefore exceeds its actual length.

Studying this phenomenon using smartphones' acceleration sensor should also be integrated in a more sophisticated instructional setting, connected with other phenomena [6, 7] addressing the students' misconceptions which could be related to this concept [8]. Anyway, current research shows that using this method to study free fall and oscillation phenomena could at least increase curiosity and motivation of the students when they learn with smartphones or tablets as experimental tools [9].

References

1. Briggle, J.: Analysis of pendulum period with an iPod touch/iPhone. Phys. Educ. **48**(3), 285–287 (2013)
2. Kuhn, J., Vogt, P.: Analyzing spring pendulum phenomena with a smartphone acceleration sensor. Phys. Teach. **50**, 504–505 (2012 Dec)
3. Vogt, P., Kuhn, J.: Acceleration sensors of smartphones: possibilities and examples of experiments for application in physics lessons. Front. Sensors. **2**(1), 1–9 (2014)
4. https://ogy.de/sparkvue
5. https://ogy.de/accelogger
6. Kuhn, J.: Relevant information about using a mobile phone acceleration sensor in physics experiments. Am. J. Phys. **82**(2), 94 (2014)
7. Vogt, P., Kuhn, J.: Analyzing collision processes with the smartphone acceleration sensor. Phys. Teach. **52**(2), 118–119 (2014)
8. Hall, J.: iBlack Box? Phys. Teach. **50**(5), 260 (2012)
9. Hochberg, K., Kuhn, J., Müller, A.: Using smartphones as experimental tools – effects on interest, curiosity and learning in physics education. J. Sci. Educ. Technol. **27**(5), 385–403 (2018)

Coupled Pendulums on a Clothesline

36

Michael Thees, Sebastian Becker-Genschow, Eva Rexigel,
Nils Cullman, and Jochen Kuhn

The importance of coupled oscillations for natural and engineering sciences is undisputed. But due to the complex processes behind the phenomena, its educational arrangement remains challenging.

Based on traditional setups of coupled spring pendulums [1–3] and torsion pendulums [3, 4], we investigate a low-cost experiment consisting of two clothespins that are mounted on a fine string [5]. We use a simple approach via spring pendulums to describe the dependencies of the coupling between the clothespins, particularly with regard to the distance between their mounting points. The theoretical approach coincides with experimental data that is extracted via mobile video analysis using tablet PCs [6] and the apps Viana [7] and Graphical Analysis [8].

M. Thees (✉) · E. Rexigel · N. Cullman
Technische Universität Kaiserslautern, Department of Physics, Physics Education Research Group,
Kaiserslautern, Germany
e-mail: theem@physik.uni-kl.de; rexigel@physik.uni-kl.de; ncullman@rhrk.uni-kl.de

S. Becker-Genschow
Digitale Bildung, Department Didaktiken der Mathematik und der Naturwissenschaften,
Universität zu Köln, Cologne, Germany
e-mail: sebastian.becker-genschow@uni-koeln.de; sbeckerg@uni-koeln.de

J. Kuhn
Ludwig-Maximilians-Universität München (LMU Munich), Faculty of Physics, Chair of Physics
Education, Munich, Germany
e-mail: jochen.kuhn@lmu.de

36.1 Theoretical Background

Compared to a traditional setup of two spring-coupled spring pendulums with identical masses, we use two identical clothespins mounted on a fine string with identical distances between the pins and the mounting of the string (Figs. 36.1 and 36.2), resulting in two string-coupled torsion pendulums.

Writing Newton's second law for x_1 and x_2 of the system in Fig. 36.1a yields

$$-kx_1 - k_{12}(x_1 - x_2) = m\ddot{x}_1 \text{ and } -kx_2 - k_{12}(x_2 - x_1) = m\ddot{x}_2,$$

which, when written in terms of the center of mass $X = \frac{1}{2}(x_1 + y_2)$ and difference coordinate, $Y = \frac{1}{2}(x_1 - y_2)$ become simple harmonic oscillator equations [3]:

$$\ddot{X} = -\frac{k}{m} \cdot X \text{ and } \ddot{Y} = \frac{k + 2k_{12}}{m} \cdot Y.$$

Thus, the system has two "normal" modes of oscillation: one with frequency squared given by $\omega^2 = k/m$, corresponding to the symmetric motion of both masses together, and one given by

$$\Omega^2 = \frac{k + 2k_{12}}{m},$$

corresponding to the anti-symmetric motion. The actual motion of each clothespin can be described as a superposition of those modes. The energy is transferred back

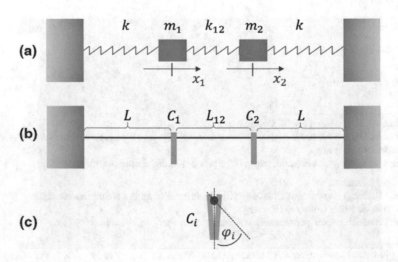

Fig. 36.1 Schematic comparison between coupled spring pendulums (**a**) and coupled torsion pendulums ([**b**] frontal view, [**c**] side view); x_i and φ_i describe the corresponding deviations from the rest positions

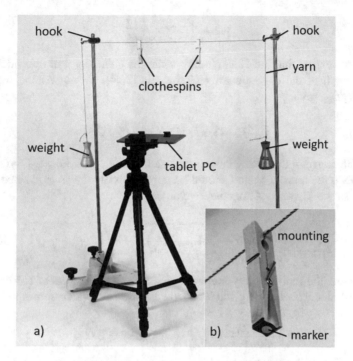

Fig. 36.2 (**a**) Photo of the experimental setup. (**b**) Detailed view of the tight mounting and one of the markers beneath the clothespins

and forth between the pins with a frequency given by half the beat frequency of the superposition, so the time it takes the energy to transfer from one pin to the other is

$$\tau_{\text{transfer}} = \frac{T_{\text{beat}}}{4} = \frac{\pi}{\Omega - \omega} = \pi \cdot \frac{\Omega + \omega}{\Omega^2 - \omega^2} = \pi \cdot \frac{\Omega + \omega}{2k_{12}/m}. \qquad (36.1)$$

Likewise, we can apply the torque equation [3] to each of the clothespins, using c and c_{12} for the Hooke's law proportionality constants of the small angular displacements of the pins from their equilibrium position, arriving at

$$-M \cdot g \cdot r \cdot \sin \varphi_1 - c\varphi_1 - c_{12}(\varphi_1 - \varphi_2) = I\ddot{\varphi}_1$$

and

$$-M \cdot g \cdot r \cdot \sin \varphi_2 - c\varphi_1 - c_{12}(\varphi_2 - \varphi_1) = I\ddot{\varphi}_2,$$

where M, r, and I are the mass, momentum arm to the center of mass, and moment of inertia of one single pin, respectively. Thus, for small angles, and with the identifications $k \rightarrow (Mgr + c)$, $k_{12} \rightarrow c_{12}$, and $m \rightarrow I$, we expect that the energy will be transferred between the clothespins in a time

$$\tau_{\text{transfer}} = \pi \cdot \frac{\Omega + \omega}{2c_{12}/I}. \tag{36.2}$$

Finally, we expect that the Hooke's law constants c and c_{12} will depend inversely on the length of the corresponding clothesline [3, 4], so we write $c = C/L$ and $c_{12} = C_{12}/L_{12}$, giving

$$\tau_{\text{transfer}} = \frac{\pi}{2} \cdot I \cdot (\Omega + \omega) \cdot L_{12}/C_{12}. \tag{36.3}$$

This result suggests that if L_{12} is about the same length as L, i.e., $L_{12} = L + x$ with $x \ll L$, then the transfer period should be approximately linear in x, since $(\Omega + \omega)$ depends so weakly on x. Altogether, we obtain

$$\tau_{\text{transfer}} = \frac{\pi}{2} \cdot \frac{I}{C_{12}} \cdot \left(\sqrt{\left(Mgr + \frac{C}{L} + 2\frac{C_{12}}{L+x} \right)/I} + \sqrt{\left(Mgr + \frac{C}{L} \right)/I} \right) \cdot (L+x).$$

Furthermore, the ratio of the y-intercept of the line to the slope of the line in this theoretical model should be given by L; the experimental value comes to

$$\frac{y - \text{intercept}}{\text{slope}} = \frac{1.45 \text{ s}}{7.37 \frac{\text{s}}{\text{m}}} = (0.197 \pm 0.006)\text{m}$$

(Fig. 36.4), a reasonably satisfying result.

36.2   Experimental Setup

The two clothespins are mounted tightly on the fine string ("the clothesline"), which is under tension with the help of two weights (we used $m = 0.5$ kg). The tablet is placed beneath the pins with its display pointing upwards in order to record a video of the pins' motion with the front camera of the device (the default recording rate is about 30 fps, iPad 4 mini). To improve tracking with the video analysis software, it is helpful to attach a colored marker beneath each clothespin (Fig. 36.2b).

To conduct the experiment, one clothespin is deflected manually and released, while the oscillation is recorded by the tablet PC. Due to the coupling via the string, the energy of the oscillation is transferred between the clothespins periodically. After several transfer periods, the recording is stopped and the video is imported into the app Viana, where the position of each marker is tracked and displayed automatically. We obtain the following diagram (Fig. 36.3) of the oscillation over time.

To determine τ_{transfer}, the data are imported into Graphical Analysis [8], where we can read out the period with the help of a precise cursor. A systematical variation of the distance between the clothespins resulted in an almost linear relation between the distance $x = L_{12} - L$ and τ_{transfer} (Fig. 36.4), in accordance to our theoretical approach (Eq. 36.4).

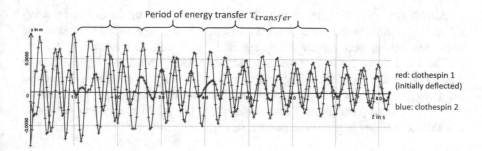

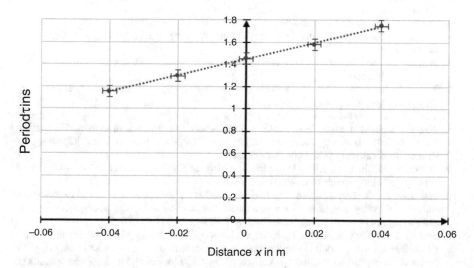

Fig. 36.3 Screenshot of the deviations over time of each clothespin, using $L_{12} = 0.2$ m and $L = 0.2$ m (the y-axis points in the direction of movement); the period of energy transfer $\tau_{transfer}$ is marked

Fig. 36.4 Period-distance-diagram for $L = 0.2$ m; a simple linear regression yields $\tau_{transfer}(x) = (7.37 \pm 0.17)\,\frac{s}{m}\cdot x + (1.45 \pm 0.01)$ s and the coefficient of determination $R^2 = 0.99$

36.3   Discussion

Combining this low-cost experiment with the simple but effective method of mobile video analysis results in a low-cost setup that can easily be used for educational purposes in school laboratories. Based on concepts of traditional spring-coupled pendulums, this case is useful for qualitative and quantitative investigations, e.g., with the presented simple theoretical approach.

Alternatively, the coupling could be varied via the tension of the string using different weights. Furthermore, the focus of the investigations could also be on the phase relationship between the oscillating clothespins, e.g., their normal modes.

The validation of our approach and further investigations are limited due to additional vibrations of the string and the pins, leading to a nonlinear impact on the energy transfer. Anyway research shows that learning with mobile video analysis can also increase conceptual understanding [9, 11] while decreasing irrelevant cognitive effort and negative emotions [10, 11].

References

1. Preyer, N.W.: The coupled harmonic oscillator: not just for seniors anymore. Phys. Teach. **34**, 52 (1996 Jan)
2. Carnevali, A., Newton, C.L.: Coupled harmonic oscillators made easy. Phys. Teach. **38**, 503 (2000 Nov)
3. Demtröder, W.: Mechanics and Thermodynamics. Springer International Publishing, 2017
4. Saxl, E.J., Allen, M.: Period of a torsion pendulum as affected by adding weights. J. Appl. Phys. **40**, 2499 (1969)
5. Hilscher, H. (Ed.), Physikalische Freihandexperimente. Band 1: Mechanik. 3. Auflage (Aulis Verlag, 2010), 228–229
6. Becker, S., Klein, P., Kuhn, J.: Video analysis on tablet computers to investigate effects of air resistance. Phys. Teach. **54**, 440–441 (2016 Oct)
7. Viana requires iOS 8.1 or higher. https://ogy.de/Viana
8. Vernier Graphical Analysis requires iOS 8.0 or higher or Android 4.0 or higher. This application is available from https://ogy.de/VernierGraphicalAnalysis (for iOS) and https://ogy.de/VernierGraphicalAnalysisAndroid (for Android)
9. Becker, S., Gößling, A., Klein, P., Kuhn, J.: Using mobile devices to enhance inquiry-based learning processes. Learn. Instr. **69**, 101350 (2020)
10. Hochberg, K., Becker, S., Louis, M., Klein, P., Kuhn, J.: Using smartphones as experimental tools—a follow-up: cognitive effects by video analysis and reduction of cognitive load by multiple representations. J. Sci. Educ. Technol. **29**(2), 303–317 (2020)
11. Becker, S., Gößling, A., Klein, P., Kuhn, J.: Investigating dynamic visualizations of multiple representations using mobile video analysis in physics lessons: effects on emotion, cognitive load and conceptual understanding. Zeitschrift für Didaktik der Naturwissenschaften. **26**(1), 123–142 (2020)

Superposition of Oscillation on the Metapendulum: Visualization of Energy Conservation with the Smartphone

37

David Weiler and Arne Bewersdorff

The Metapendulum is a combination of both the simple gravity pendulum and the spring pendulum. Both excite each another and, therefore, can be used as an example of mechanical resonance phenomena. The data of a smartphone's accelerometer can be used to record the acceleration of the pendulum during the oscillation movement. With that, the resonance within the system can be visualized.

37.1 Theoretical Background

The use of the smartphone to record the acceleration of the spring pendulum (Chap. 32) [1] as well as the gravity pendulum (Chap. 29) [2] and coupled pendulums (Chap. 36) [3] has already been described in this book. The Metapendulum [4] (also known as "Gorelikpendulum") now combines the linear elastic oscillation of a spring pendulum with the harmonious oscillation of a simple gravity pendulum.

In the experimental setup, a cord with a smartphone as pendulum mass is attached to a spring pendulum (Figs. 37.1 and 37.2). The vertical oscillation of the spring pendulum and the horizontal oscillation of the gravity pendulum alternate periodically. This is due to the fact that the resonance frequency of the two partial pendulums is calculated to be identical. One pendulum can excite the other and vice versa. The energy fluctuates between kinetic energy and potential energy in the spring pendulum, and kinetic and potential energy in the gravity pendulum. The motion can be characterized into three states (see Fig. 37.3, and video [5]):

D. Weiler (✉)
Eberhard Karls University of Tübingen, Tübingen, Germany
e-mail: david-christoph.weiler@uni-tuebingen.de

A. Bewersdorff
Technical University of Munich, Munich, Germany

J. Kuhn, P. Vogt (eds.), *Smartphones as Mobile Minilabs in Physics*,
https://doi.org/10.1007/978-3-030-94044-7_37

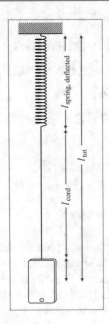

Fig. 37.1 Setup of the Metapendulum

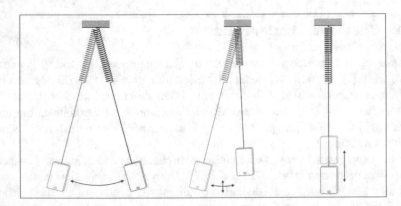

Fig. 37.2 Exemplary experimental setup of the Metapendulum with smartphone

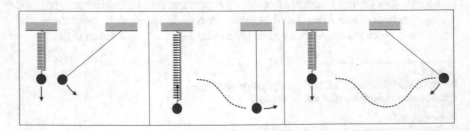

Fig. 37.3 The three states of motion of the Metapendulum

1. The total energy of the system is inherent in the horizontal pendulum movement of the cord. It is distributed between the potential energy (velocity is zero) and the kinetic energy (maximum velocity at rest position).
2. The energy is divided between the two modes of oscillation. This causes both the stretching of the spring and the horizontal deflection of the thread pendulum.
3. The total energy is stored in the vertical pendulum movement of the spring pendulum. Here the forms of the maximum potential energy (upper reversal point), the maximum kinetic energy (passage through the rest position), and the maximum tension energy (lower reversal point) can be distinguished.

In order to achieve a cyclic shift between the pendulum oscillation and the spring oscillation, the length of the cord must match with the spring constant of the coil spring.

First, the mass m of the smartphone with the smartphone case and the period T of the spring pendulum must be determined by weighing and measuring. Therefore, the spring constant D is calculated with the formula

$$T_{\text{spring}} = 2\pi \sqrt{\frac{m}{D}}. \tag{37.1}$$

The transformation to D gives

$$D = \frac{4\pi^2 m}{T^2}. \tag{37.2}$$

Now it must be calculated how long the combined length of the spring pendulum, the cord, and the smartphone have to be in order to achieve the intended resonance phenomena. For this to happen, the period of the gravity pendulum has to be twice the period of the spring pendulum:

$$T_{\text{gravity}} = 2 \cdot T_{\text{spring}}. \tag{37.3}$$

This is due to the fact that during one period of oscillation of the gravity pendulum, the projection of the movement of the mass has twice the period of a spring pendulum (Fig. 37.4). Under these circumstances, resonance occurs.

Let the formula for the period of the gravity pendulum be

$$T_{\text{gravity}} = 2\pi \sqrt{\frac{l}{g}}. \tag{37.4}$$

The total length of the pendulum consists of the length of the spring pendulum, the cord length (gravity pendulum), and the distance to the center of mass of the smartphone (usually half the length of the smartphone). Resolving Eq. (37.4) with Eqs. (37.3) and (37.1), the length is determined to

Fig. 37.4 One oscillation of
the spring pendulum
corresponds to half an
oscillation of the gravity
pendulum

$$l_{\text{total}} = 4\frac{m \cdot g}{D}. \tag{37.5}$$

To determine the cord length, the length of the deflected spring and half the length of
the smartphone have to be subtracted from the total length.

The length of the deflected spring can be calculated with the help of Hooke's law
using the formula

$$l_{\text{spring,deflected}} = \Delta l + l_{\text{spring}} = \frac{m \cdot g}{D} + l_{\text{spring}}. \tag{37.6}$$

Alternatively, it can also be measured directly by determining the spring constants.

37.2 Analysis

During the experiment, the x-, y-, and z-axis acceleration is recorded from the sensors
of the smartphone with the app Phyphox over time—minus the acceleration due to
gravity (Fig. 37.5). Data can be exported to an Excel sheet. A start delay of 3 s or
starting the measurement via Bluetooth both have proven to be effective.

Fig. 37.5 Recording the acceleration (x,y,z) in Phyphox

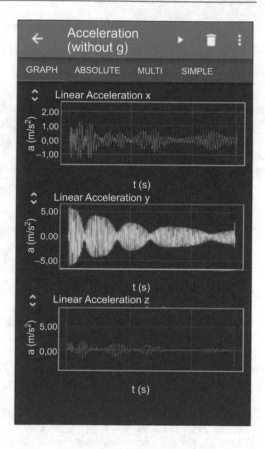

Fig. 37.6 Horizontal acceleration must be calculated by adding x- and z-components

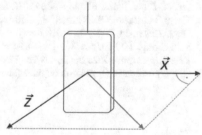

For the analysis, noise at the beginning and at the end of the experiment, which are caused by releasing and stopping the smartphone, is removed. The values of the y-axis sensor can be used for the vertical acceleration. Since the smartphone rotates during the experiment, the Pythagorean theorem must be used to add the acceleration of the horizontal movement from the x- and z-axis sensor data (Fig. 37.6).

The two pendulum accelerations over time are compared in the diagram seen in Fig. 37.7. Transitions between vertical and horizontal acceleration are clearly recognizable in this diagram. Oscillation decreases due to friction losses until it finally

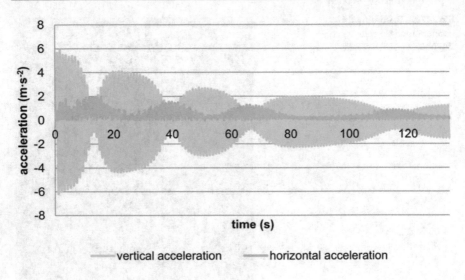

Fig. 37.7 Exemplary analysis with Excel

comes to a complete standstill. Thus, the Metapendulum can be used to exemplarily explain resonance phenomena in coupled modes.

References

1. Kuhn, J., Vogt, P.: Analyzing spring pendulum phenomena with a smartphone acceleration sensor. *Phys. Teach.* **50**, 504–505 (2012 Nov)
2. Vogt, P., Kuhn, J.: Analyzing simple pendulum phenomena with a smartphone acceleration sensor. *Phys. Teach.* **50**, 439–440 (2012 Oct)
3. Thees, M., Becker, S., Rexigel, E., Cullman, N., Kuhn, J.: Coupled pendulums on a clothesline. *Phys. Teach.* **56**, 404–405 (2018 Sept)
4. Schlichting, H.-J., Ucke, C.: Das 'Metapendel' oder: Eine sich selbst antreibende Schaukel. *Physik in Unserer Zeit.* **26**(1), 41–42 (1995)
5. "Superposition of Oscillation on the Metapendulum," YouTube, https://youtu.be/Oyr7sBzs4cY

Demonstration of the Parallel Axis Theorem Through a Smartphone

I. Salinas, M. H. Gimenez, J. A. Monsoriu, and J. A. Sans

New learning strategies try to extend the use of common devices among students in physics lab practices. In particular, there is a recent trend to explore the possibilities of using smartphone sensors to describe physics phenomena (Chap. 62) [1, 2]. On the other hand, the study of the moment of inertia by the use of the torsion pendulum is a typical example in the first courses of fundamentals of physics [3]. This example allows the exploration of harmonic motion, Newton's second law, the moment of inertia theory, and the parallel axis theorem all in one. Here, we report the use of the accelerometer sensor of a smartphone to visualize and demonstrate the parallel axis theorem in a torsion pendulum.

38.1 Experiment

The study of the torsion pendulum motion will serve to visualize forced, damped, or simple harmonic motion and explore the implications of the moment of inertia and the parallel axis theorem [4, 5].

In the experiment described in Fig. 38.1, a rod rotates around its rotation axis subjected to the force of a spring (stiffness k). In order to measure the acceleration caused in the extreme of the rod, we took advantage of the accelerometer sensor of the smartphone, whereas a known mass m is placed at several distances from the center of the rod. Thus, applying Newton's second law, we know that the moment of force or torque M_T is related to the angular acceleration α and the moment of inertia I. Therefore, the motion is dominated by the stiffness of the spring and the angle shifted θ with the following relation:

I. Salinas (✉) · M. H. Gimenez · J. A. Monsoriu · J. A. Sans
Universitat Politècnica de València, València, Spain
e-mail: isalinas@fis.upv.es; mhgimene@fis.upv.es; jmonsori@fis.upv.es; juasant2@upv.es

Fig. 38.1 Scheme of the experiment performed

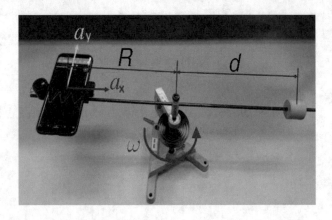

$$M_T = I\alpha = -k\theta. \tag{38.1}$$

Then, the equation that must be solved is:

$$\ddot{\theta} + \frac{k}{I}\theta = 0, \tag{38.2}$$

where the angular frequency is:

$$\Omega = \sqrt{\frac{k}{I}} = \frac{2\pi}{T}. \tag{38.3}$$

The moment of inertia of the system can be described as the addition of the moment of inertia of its components, as follows:

$$I = I_{rod} + I_{smartphone} + I_{mass}. \tag{38.4}$$

Now, applying the parallel axis (or Huygens-Steiner) theorem to m that says the moment of inertia around any axis (I_{mass}) separated a distance d to the center of mass can be obtained from the moment of inertia of a parallel axis passing through the center of mass of the object (I_{mass0}),

$$I_{mass} = I_{mass0} + md^2. \tag{38.5}$$

Then, the moment of inertia of the system can be expressed as a function of the distance, such as

$$I = I_{rod} + I_{smartphone} + I_{mass0} + md^2 = I_0 + md^2. \tag{38.6}$$

The period squared of the oscillation can be expressed as

$$T^2 = \frac{4\pi^2 I}{k} = \frac{4\pi^2 I_0}{k} + \frac{4\pi^2 m}{k} d^2. \tag{38.7}$$

In summary, there is a direct relation between the period of the oscillation registered in the extreme of the rod and the distance of the mass to its center of gravity.

38.2 Analysis and Discussion

By means of a slight shift of the rod in the spring, the system starts to oscillate. The oscillations of the system are collected by the Android application Physics Toolbox Suite [6], taking advantage of the smartphone's accelerometer sensor. The representation of the oscillations is shown in Fig. 38.2, which allows us to calculate the period (T) of each oscillation for the normal and tangential component of the acceleration.

The values of the time period of the tangential component of the acceleration at each position of the mass m of 0.236 kg are represented in Table 38.1.

The square of the values showed in Table 38.1 are displayed in Fig. 38.3, where one can observe a linear trend, according to Eq. (38.7). The high quality of the fit $(r^2 = 0.9999)$ offered a large reliability of the parameters obtained. Thus, from the fit to a linear dependence through minimum squares and comparing with Eq. (38.7), we can obtain that the slope a_{slope} is related to the stiffness of the spring as $a_{slope} = 4\pi^2 m/k$.

Fig. 38.2 Plot of the normal and tangential components of the acceleration in function of time placing the mass m at $d = 0.20$ m

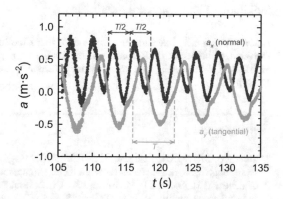

Table 38.1 Average time period of the tangential acceleration for each position (d) of the mass m

d (m)	$T_{average}$ (s)
0.05	4.87
0.10	5.18
0.15	5.65
0.20	6.24
0.25	6.92
0.30	7.68

Fig. 38.3 Linear fit of period squared in function of the square of the distance between the mass and its center of gravity

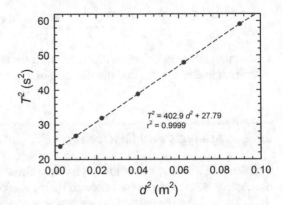

Applying Eq. (38.1), one can also find the value of the stiffness of the spring by the application of a force at a determined distance. In our case, we have applied 0.25 N at a distance of 0.3 m and we have observed that the rod rotated π rad, then $k = |M_T|/\theta = 0.0239$ Nm. The stiffness calculated from Eq. (38.7) is $k_{measured} = 4\pi^2 m/a = 0.0231$ Nm. The discrepancy obtained in the calculation of the stiffness of the spring by both of these methods is less than 4%.

In summary, we present a new way to calculate the stiffness of a spring applying previously acquired knowledge about moment of inertia and parallel axis theorem thanks to the use of the accelerometer sensor of a smartphone. This new method has been proved as an invaluable tool to bring physics experimentation to the students and discover the potential possibilities of this common device as a sensor in a multitude of basic physics experiments.

Acknowledgments The authors would like to thank the Institute of Educational Sciences of the Universitat Politècnica de València (Spain) for the support of the Teaching Innovation Groups MoMa and e-MACAFI. JAS acknowledges Ramón y Cajal fellowship program (RYC-2015-17482).

References

1. Monteiro, M., Stari, C., Cabeza, C., Martí, A.C.: Magnetic field 'flyby' measurement using a smartphone's magnetometer and accelerometer simultaneously. Phys. Teach. **55**, 580 (2017 Dec)
2. Giménez, M.H., Salinas, I., Monsoriu, J.A.: Direct visualization of mechanical beats by means of an oscillating smartphone. Phys. Teach. **55**, 424 (2017 Oct)
3. Green, R.: Calibrated torsion pendulum for moment of inertia measurements. Am. J. Phys. **26**, 498 (1958 July)
4. Boyd, J.N., Raychowdhury, P.N.: Parallel axis theorem. Phys. Teach. **23**, 486 (1985 Nov)
5. Christie, D.: Tennis rackets and the parallel axis theorem. Phys. Teach. **52**, 208 (2014 April)
6. https://ogy.de/physicstoolboxsensorsuite

Rotational Energy in a Physical Pendulum

39

Martín Monteiro, Cecilia Cabeza, and Arturo C. Marti

Smartphone usage has expanded dramatically in recent years worldwide. This revolution also has impact in undergraduate laboratories where different experiences are facilitated by the use of the sensors usually included in these devices. Recently, in several articles published in the literature [1, 2], the use of smartphones has been proposed for several physics experiments. Although most previous articles focused on mechanical experiments, an aspect that has received less attention is the use of rotation sensors or gyroscopes. Indeed, the use of these sensors paves the way for new experiments enabling the measurement of angular velocities. In a earlier chapter of this book the conservation of the angular momentum is considered using rotation sensors (Chap. 22) [3]. In this work we present an analysis of the rotational energy of a physical pendulum.

39.1 Experimental Setup

The experimental setup consists of a smartphone on the periphery of a bicycle wheel that can rotate freely in a vertical plane as shown in Fig. 39.1. The smartphone is an LG Optimus P990 2X (three-axis gyroscope MPU3050 Invensense, accuracy 0.0001 rad/s). The moment of inertia of the wheel, easily obtained by means of small oscillations, is 0.040 kg·m^2. The distance from the center of mass of the wheel to the center of mass of the smartphone is $R = 0.30$ m and the mass of the smartphone $m = 0.14$ kg.

M. Monteiro (✉)
Universidad ORT Uruguay, Montevideo, Uruguay
e-mail: monteiro@ort.edu.uy

C. Cabeza · A. C. Marti
Universidad de la República, Montevideo, Uruguay
e-mail: cecilia@fisica.edu.uy; marti@fisica.edu.uy

© The Author(s), under exclusive license to Springer Nature Switzerland AG 2022
J. Kuhn, P. Vogt (eds.), *Smartphones as Mobile Minilabs in Physics*,
https://doi.org/10.1007/978-3-030-94044-7_39

Fig. 39.1 Smartphone mounted on a bike wheel and a scheme indicating the axes orientation

The application AndroSensor [4] running under Android was used to record the values measured. The magnitudes relevant in this work are those reported by the rotation sensor according to the x-axis. Once recorded, data can be exported and analyzed using appropriate software.

39.2   Analysis of the Motion

In our experiment the bike wheel is set in motion performing full rotations in one direction. During a single rotation the energy is very nearly conserved, and upon using the bottom of the bicycle wheel as the zero of potential energy and measuring the angle from this lowest potential energy position, we can write the total energy of the system during the rotation as

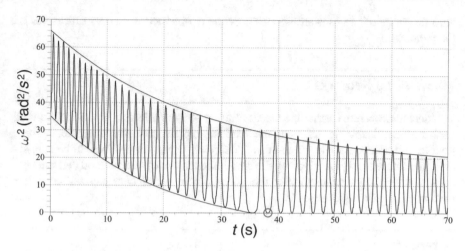

Fig. 39.2 Angular velocity squared (black), proportional to the kinetic energy, and two exponential curves fitting relative maxima and minima (red). The green circle indicates the transition from rotations to oscillations

$$E = \frac{1}{2}I\omega_{\text{bottom}}^2 = \frac{1}{2}I\omega_{\text{top}}^2 + 2mgR,$$

where I is the moment of inertia of the wheel-phone combination about the pivot point and ω_{bottom} and ω_{top} are the angular velocities when the smartphone is at the bottom or top position, respectively. Thus, conservation implies that

$$\omega_{\text{bottom}}^2 - \omega_{\text{top}}^2 = \frac{4mgR}{I} = \text{constant}.$$

In our setup the constant was determined from the parameters to be 31.3 (rad/s) [2].

As time goes by, due to the effect of a weak dissipation, the energy decreases and, at a given point, the angular velocity first vanishes. Then, the wheel reverses the direction of spinning and starts to oscillate around the stable equilibrium point. In Fig. 39.2, the square of the angular velocity, proportional to the kinetic energy, is plotted as a function of time. At the beginning of the graph, the wheel is rotating in one direction; however, at $t \approx 38$ s, indicated by the green circle, the angular velocity vanishes for the first time, and the wheel starts oscillating. Before the green circle, when the wheel is rotating, relative maxima and minima are achieved when the smartphone is passing through the lowest or highest points, respectively. After the green circle, when the wheel is oscillating, relative minima coincide with turning points or zeros of the angular velocity, while relative maxima correspond to the smartphone passing through the lowest point in one direction or the other according to the sign of the angular velocity.

In this figure, two exponential curves, fitting the relative maxima and minima, reveal that the energy decreases roughly exponentially. The vertical difference between these curves is about 31 (rad/s) [2]. This value shows excellent agreement with the value of the constant $4mgR/I$ previously obtained from direct measures of the parameters.

39.3   Final Remarks

To conclude, we remark that the use of rotation sensors allows a broad spectrum of measures applicable in different mechanical experiments, for example, spring, double, or torsion pendula, or coupled oscillators. Other possibilities, to be considered in a future work, are given by the simultaneous use of acceleration and rotation sensors that allows, among other things, to obtain a full characterization of the motion of simple systems.

References

1. See for example, Jochen Kuhn and Patrik Vogt, "Smartphones as experimental tools: Different methods to determine the gravitational acceleration in classroom physics by using every Figure 1 (a). Photographs of timer assembly. day devices," Eur. J. Phys. Educ. **4**, 16 (2013).
2. Vogt, P., Kuhn, J., Müller, S.: Experiments using cell phones in physics classroom education: The computer-aided g determination. *Phys. Teach.* **49**, 383 (2011 Sept)
3. Shakur, A., Sinatra, T.: Angular momentum. *Phys. Teach.* **51**, 564 (2013 Dec)
4. Several applications available at http://play.google.com allow recording the values measured by the sensors.

Part VII

Acoustical Logging and the Speed of Sound

Determining the Speed of Sound with Stereo Headphones

Patrik Vogt and Jochen Kuhn

In this example we describe how the speed of sound can be determined using simple stereo headphones (ear buds) and sound analysis freeware.

When watching the start of a race from a distance of several hundred meters, one notices a time difference Δt between the visual perception of the start signal and hearing the bang. Sound is therefore not propagated instantaneously, but at a finite speed c. There are numerous other examples in the daily lives of pupils that support this conclusion. For instance, the observed time difference between the perception of thunder and lightning is a particularly impressive example that enables us to calculate how far away a storm is.

There are two main methods of determining the speed of sound in air in experiments in school instruction. The first method involves producing standing waves in a glass tube [1]. The location of nodes and antinodes can be found and the speed of sound calculated by multiplying wavelength and frequency. Alternatively, the so-called transit time method can be applied: This entails calculating the quotient of the measured distance of the observer of the sound source and the time difference Δt [2]. This is without a doubt the clearer alternative for pupils, in particular for those at lower secondary school level.

If the time transit method is performed in a physics laboratory, the short distance between the observer and the sound source makes it hard to measure the time interval. For this reason a computer-aided data measurement system is usually used. However, this has two disadvantages: first, a large number of schools do not

P. Vogt (✉)
Institute of Teacher Training (ILF) Mainz, Mainz, Germany
e-mail: vogt@ilf.bildung-rp.de

J. Kuhn
Ludwig-Maximilians-Universität München (LMU Munich), Faculty of Physics, Chair of Physics Education, Munich, Germany
e-mail: jochen.kuhn@lmu.de

have a system of this kind; second, the experiment conducted conventionally can only be carried out as a demonstration, owing to cost restraints.

40.1 Determining the Speed of Sound with a Sound Card and a PC

One possibility to perform the experiment without an expensive data measurement system, but with the help of a PC equipped with a sound card, two microphones, and audio analysis software (e.g., Audacity [3]), is described in Ref. [4] and [5]. The microphones have to be set up at a distance of approximately one meter and plugged into the stereo input of the sound card (line-in) using a *y*-cable (Fig. 40.1).

After having activated the record symbol of the analysis software, a short loud acoustic impulse is produced from a location that is on a straight line with the microphones. It can, for example, be generated by a loud hand clap. When performing the stereo recording, the audio analysis software is able to analyze the two microphones separately; as a result, the sound impulse is displayed first on the microphone situated closer to the sound source and reaches the second microphone after a time interval Δt. This time interval can be measured by the audio analysis software (see below), and the velocity c at which the sound impulse moved from microphone 1 to microphone 2—i.e., the speed of sound—can be calculated by dividing the distance Δs of the two microphones by the time interval Δt:

$$c = \frac{\Delta s}{\Delta t}.$$

40.2 Low-Cost Alternative Using Headphones

If the microphones are replaced by a standard pair of headphones (Fig. 40.2), it can then be considered a low-cost experiment and can also be conducted by pupils.

Figure 40.3 shows a measurement example for headphone capsules with a 0.50-m distance. The experiment was conducted identically to the experiment described above with the two microphones. With a time difference of 1.49 ms, the speed of sound in air was calculated to be approximately 336 m/s. A comparison with the known value (343 m/s at 20 °C) demonstrates that the speed of sound in air can be

Fig. 40.1 Determining the speed of sound using the transit time method

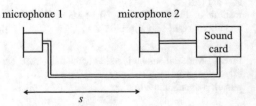

Fig. 40.2 Determining the speed of sound using headphones

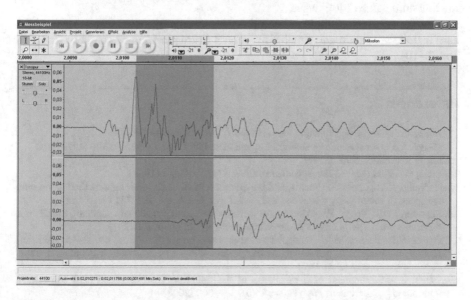

Fig. 40.3 Measurement example for a distance of 0.5 m, recorded with Audacity software

measured sufficiently accurately using an everyday object that is available to all pupils and easy to use. The literature value is situated within the error interval of the measurement [$c = (336 \pm 11)$ m/s].

As well as determining the speed of sound in air, the experiment presented in Ref. [4] can be conducted to calculate the speed of propagation of sound in water, also using a standard pair of headphones. A possible measuring distance could, for

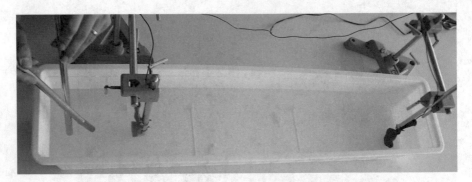

Fig. 40.4 Experiment setup to measure the speed of sound in water

example, be within a window box filled with tap water (Fig. 40.4). In order to use the headphones as hydrophones, they need to be covered in latex balloons. With a 0.57-m microphone distance and a measured time difference of 0.40 ms, the speed of sound in water is calculated to be $v \approx 1400$ m/s. This value also matches well with the literature value (1400 m/s at 5 °C).

A series of other examples of computer-aided data measurement using a sound card that are appealing and suitable for school instruction can be found in Refs. [4] and [6–10].

References

1. Lührs, O.: Gases reveal standing waves in tubes. *Phys. Educ.* **39**, 333 (2004 July)
2. Girard, J.E.: Direct measurement of the speed of sound. *Phys. Teach.* **17**, 393 (1979 Sept)
3. https://www.audacityteam.org/
4. Stein, W.: *Versuche mit der Soundkarte*. Ernst Klett, Stuttgart (1999)
5. Carvalho, C.C., Lopes dos Santos, J.M.B., Marques, M.B.: A time-of-flight method to measure the speed of sound using a stereo sound card. *Phys. Teach.* **46**, 428–431 (2008 Oct)
6. Schwarz, O., Vogt, P.: Akustische Messungen an springenden Bällen. Praxis der Naturwissenschaften – Physik in der Schule. **3/53**, 22–25 (June 2004)
7. Aguiar, C.E., Pereira, M.M.: Using the sound card as a timer. *Phys. Teach.* **49**, 33–35 (2011 Jan)
8. Hassan, U., Pervaiz, S., Anwar, M.S.: Inexpensive data acquisition with a sound card. *Phys. Teach.* **49**, 537–539 (2011 Dec)
9. Ganci, S.: Measurement of g by means of the 'improper' use of sound card software: a multipurpose experiment. *Phys. Educ.* **43**(3), 297–300 (2008)
10. White, J.A., Medina, A., Román, F.L., Velasco, S.: A measurement of g listening to falling balls. *Phys. Teach.* **45**, 175–177 (2007 March)

Stationary Waves in Tubes and the Speed of Sound

41

Lutz Kasper, Patrik Vogt, and Christine Strohmeyer

The opportunity to plot oscillograms and frequency spectra with smartphones creates many options for experiments in acoustics, including several ones in this book (Chaps. 2, 48 and 49) [1–3]. The activities presented in this chapter are intended to complement these applications, and include an approach to determine sound velocity in air by using standard drain pipes [4, 5] and an outline of an investigation of the temperature dependency of the speed of sound.

Sound Propagation in Drain Pipes "Intermateable" PVC drain pipes (e.g., length $L = 3$ m, threaded or simply belled on one end, diameter $d = 6–16$ cm), purchased at hardware stores often at less than a dollar (U.S.) per foot, are used to set up a sound tube as long as possible (minimum 5 m).[1] If a student whispers into one end of the tube, despite the large distance covered, students at the other end will hear it well (Fig. 41.1). Since the sound spreads primarily in the direction of the tube, and not in three dimensions, the sound at the other end will exceed the auditory threshold. The "whisper experiment" will be especially impressive if the sound tube is led around one or more corners, which can simply be realized with elbow fittings (Fig. 41.2).

In addition, another phenomenon can be observed: the acoustical signal propagating in the tube is reflected at the open tube end and is perceived again as

---

[1] A development of our experiment (with a tube length below 50 cm) is described in Ref. [5].

L. Kasper (✉)
Department of Physics, University of Education Schwäbisch Gmünd, Schwäbisch Gmünd, Germany
e-mail: lutz.kasper@ph-gmuend.de

P. Vogt
Institute of Teacher Training (ILF) Mainz, Mainz, Germany
e-mail: vogt@ilf.bildung-rp.de

C. Strohmeyer
University of Education, Freiburg, Germany

Fig. 41.1 A girl listens to the voice of a classmate who is positioned at the other end of the long pipe

Fig. 41.2 Sound tube setup with "intermateable" drain pipes

an echo with a slight delay. A loud, abrupt acoustical signal, e.g., a snap of the fingers, may even be heard several times because of multiple reflections from the ends of the tube. With every echo the propagation distance rises by double the length of the tube. If the acoustical signals are recorded with a smartphone oscilloscope app (Fig. 41.3), the sound velocity can be calculated with high accuracy from the readable time differences.

Fig. 41.3 Illustration of the distance covered by the acoustic signal (a finger "snap") during the evaluation of the third echo

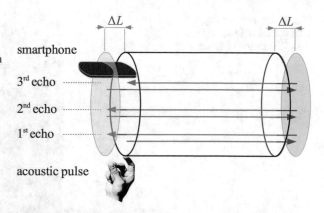

Fig. 41.4 Screenshot from the oscilloscope app; the first amplitude represents the original snapping noise, the following amplitudes are caused by following echoes

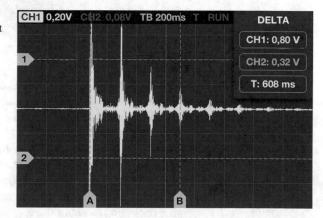

End-Pipe Correction For more accurate results, the end-pipe correction has to be taken into account, which indicates that the reflection point of the sound wave (its pressure node) is located slightly outside the pipe. The end-pipe correction ΔL depends on the radius R of the pipe, as the following applies [6]:

$$\Delta L \approx 0.61R.$$

An easy way to determine the end-pipe correction in a classroom experiment is described in Ref. [7].

The exemplary measurement was realized with a sound tube of length $L = 34.4$ m and radius $R = 8.1$ cm; the room temperature was 18 °C. As can be seen in Fig. 41.4, the transit time of the third echo is 0.608 s. With $\Delta L = 0.61R$, the sound velocity c is given by

$$c = \frac{\Delta s}{\Delta t} = \frac{6L + 10 \cdot 0.61R}{\Delta t} \approx 340 \ \frac{\text{m}}{\text{s}}.$$

Fig. 41.5 Investigating the temperature dependency of the speed of sound in air

The value matches well with the value given in the literature (342 m/s at 18 °C) [8]. Since the smartphone microphone was located at the place where the sound was generated (same pipe end) and used to measure the third echo, the end-pipe correction will be traversed ten times altogether (Fig. 41.3).

In order to minimize the error in measurement, the time difference is analyzed from the original snapping sound to an echo as late as possible instead of using the first echo, resulting in a relative error of about 0.5%. Depending on pipe length and radius, the end-pipe correction can be neglected in many circumstances—in the example above, the end-pipe correction is about 0.2%.

Determining the Tube Length Instead of determining the sound velocity, the described setup can be used to experimentally determine the tube length. For the sound tube used in Fig. 41.2 ($R = 3.5$ cm), the transit time between the original signal and the third echo was 199 ms. Therefore the tube length is acoustically determined to be 11.3 m (at a temperature of 18 °C). The length measured with scale was 11.25 m, and the discrepancy is readily accounted for by the multiple elbow bends.

Temperature Dependency of the Speed of Sound The relatively long propagation distances ensure that the error in the determined sound velocity is very small. Thus, the temperature dependency of sound velocity in air may be verified with the same setup. A sound tube similar to the one in Fig. 41.5 was set up in winter, first outside and then inside the building. In order to facilitate the conventional measurement of tube length, the sound tube was set up without elbow bends. The measured values matched reasonably well with the theoretically anticipated values [8] (see Table 41.1)—the deviations are in the direction one would expect if one were to include the effect of the water vapor in the air.

Table 41.1 Temperature dependency of the speed of sound

	Outdoor measurement	Indoor measurement
Pipe length	9.48 m	9.48 m
Pipe radius	0.07 m	0.07 m
Temperature	0 °C	24 °C
Time difference between signal and the echo	0.229 s	0.219 s
Speed of sound in the experiment	333 m/s	348 m/s
Speed of sound in dry air (literature value)	331 m/s	346 m/s

References

1. Kuhn, J., Vogt, P.: Analyzing acoustic phenomena with a smartphone microphone. Phys. Teach. **50**, 182–183 (2013)
2. Kuhn, J., Vogt, P., Hirth, M.: Analyzing the acoustic beat with mobile devices. Phys. Teach. **52**, 248–249 (2014)
3. Vogt, P., Kuhn, J., Neuschwander, D.: Determining ball velocities with smartphones. Phys. Teach. **52**, 309–310 (2014)
4. Vogt, P., Kasper, L.: Bestimmung der Schallgeschwindigkeit mit Smartphone und Schallrohr. Unterricht Physik. **140**, 43–44 (2014)
5. Hirth, M., Gröber, S., Kuhn, J., Müller, A.: Experimentelle Untersuchung akustischer Resonanzen in eindimensionalen Wellenträgern mit Smartphone und Tablet-PC. Physik. **1**(14), 12–25 (2015)
6. Levine, H., Schwinger, J.: On the radiation of sound from an unflanged circular pipe. Phys. Rev. **73**, 383 (1948)
7. Ruiz, M.J.: Boomwhackers and end-pipe corrections. Phys. Teach. **52**, 73–75 (2014)
8. Online calculator for the temperature dependency of sound velocity in air. https://ogy.de/speedofsound

Tunnel Pressure Waves: A Smartphone Inquiry on Rail Travel

42

Andreas Müller, Michael Hirth, and Jochen Kuhn

When traveling by rail, you might have experienced the following phenomenon: The train enters a tunnel, and after some seconds a noticeable pressure change occurs, as perceived by your ears or even by a rapid wobbling of the train windows. The basic physics is that pressure waves created by the train travel down the tunnel, are reflected at its other end, and travel back until they meet the train again. Here we will show (i) how this effect can be well understood as a kind of large-scale outdoor case of a textbook paradigm (Chap. 41 and [1–4]), and (ii) how, e.g., a prediction of the tunnel length from the inside of a moving train on the basis of this model can be validated by means of a mobile phone measurement.

42.1 Model

When a pressure wave runs down a tube (a sound wave duct, scientifically speaking) and hits the end, it (more precisely, part of its intensity) will be reflected. We will refer to the two ends of the tunnel as the train entrance and exit. If, as in the case of a tunnel, the tube has open ends (low acoustic impedance of the surrounding medium), a phase inversion occurs, and compression (C) will be reflected as rarefication

A. Müller (✉)
Faculty of Science/Physics Section and Institute of Teacher Education, University of Geneva, Geneva, Switzerland
e-mail: Andreas.Mueller@unige.ch

M. Hirth
Dr. Carl-Hermann-Gymnasium, Schönebeck, Germany
e-mail: hirth@gym-hermann.bildung-lsa.de

J. Kuhn
Faculty of Physics, Chair of Physics Education, Ludwig-Maximilians-Universität München (LMU Munich), Munich, Germany
e-mail: jochen.kuhn@lmu.de

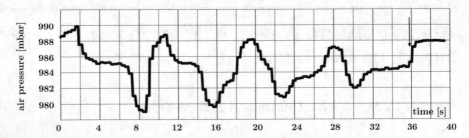

Fig. 42.1 Pressure variation as function of time during a train passage through a tunnel. The entrance moment of the passenger is at $t = 0$, and the exit moment is indicated by the black arrow, in between a series of pressure peaks and dips (see text for detailed discussion)

(R) and vice versa [1, 2]. With multiple reflections a sequence of rarefication and compression pulses RCRCRC. . . will hit the train and the ear of a passenger. Qualitatively, this is precisely what can be seen from Fig. 42.1, showing the pressure variation as a function of time during a train passage through a tunnel, as measured by the pressure sensor of a smartphone (see below for experimental details). In order to validate our basic model quantitatively, we consider the time difference Δt between two consecutive R pulses, i.e., the dips in Fig. 42.1. This time difference is given by

$$\Delta t = \frac{2l}{v_s + v_t},\qquad (42.1)$$

with l, v_t, v_s being the length of the tunnel, and the speed of the train and of sound, respectively. Throughout this contribution, we will assume v_t to be constant while the train traverses the tunnel. Equation (42.1) can be obtained as follows: In the sequence RCRCRC. . . initiated by the front end wave of the train, the R pulses arrive after an uneven number ($u = 1, 3, 5. . .$) of reflections. The time of arrival of the R pulse after u reflections is given by

$$t_u = t_0 + \Delta t_u,\qquad (42.2)$$

where t_0 is the build-up time of the initial pulse after the train has entered the tunnel, and Δt_u the time of propagation from this moment until the R pulse after u reflections hits the train again. This propagation time is given by the speed of sound and by the total distance it has to cover, with the following contributions: $l - t_0 v_t$ is the first stretch from the origin of the pulse to the exit; then before the uth reflection at exit, the sound has to cover $(u - 1)l$ times the full tunnel length from exit to entrance and back to exit; finally $(l - t_u v_t)$ for the last stretch from the last reflection at exit back to the train, which in between has covered the distance $t_u v_t$ in the tunnel. Putting this together yields

$$\Delta t_u = v_S^{-1}[(l - t_0 v t) + (u - 1)l + (l - t_u v_t)]$$
$$= v_S^{-1}[(u + 1)l - t_0 v_t - t_u v_t)]. \tag{42.3}$$

Inserting this in Eq. (42.2) and solving for t_u, we obtain

$$t_u = \frac{t_0(v_s - v_t)}{v_s + v_t} + (u + 1)\frac{l}{v_s + v_t}. \tag{42.4}$$

Inserting in $\Delta t = t_{u+2} - t_u$ finally leads to Eq. (42.1). Note that in the above considerations, t_0 is determined by a complicated process of 3-D fluid dynamics (depending for example on the shapes of the train and tunnel), which would be impossible to calculate without advanced numerical methods; Δt_u is determined by the 1-D dynamics of sound propagation and reflection in tubes and is accessible to high school physics. Considering the differences in Eq. (42.1) allows one to avoid a theoretical treatment of the very difficult t_0 part.

The result [1] for Δt can also be used to infer the other quantities, e.g., the tunnel length (see next section):

$$l = \frac{1}{2}\Delta t(v_s + v_t). \tag{42.5}$$

A form of this relationship for the practically important case where v_t is not known, but only the time of passage of the train through the tunnel (t_p), is obtained by inserting $v_t = l/t_p$ in Eq. (42.5) and solving for l:

$$l = \frac{1}{2}\left(1 - \frac{1}{2}\Delta t/t_p\right)^{-1}\Delta t v_s. \tag{42.5*}$$

We now turn to an experimental validation of the above results based on smartphone measurements.

42.2 Experiment

The measurements were made with the integrated pressure sensor of a Samsung Galaxy S4 smartphone. It was used in combination with the app Androsensor, which has a good degree of user friendliness, allowing the readout of all built-in sensors and data-logging to a csv file [5]. The maximal readout frequency is 4 s^{-1}, and reliability at least for relative pressure measurements is satisfactory [6].

The data of Fig. 42.1 were taken from the Heiligenberg Tunnel in the state of Rhineland Palatinate of Germany (49°26′N, 7°51–52′E), with parameters as follows:

$$l(\text{tunnel lenght}) = 1347 \text{ m (official value)}$$
$$v_t(\text{train velocity}) = 38 \text{ m/s (measurement, see below)} \qquad (42.6)$$
$$v_s(\text{sound velicity}) = 343 \text{ m/s (tabulated value for 20 °C)}$$

The measurement begins exactly when the train enters the tunnel and ends shortly after, and the sequence of main observations is as follows. At the moment of entrance, there is a slight pressure increase with respect to the value outside the tunnel (not shown). Then occurs the sequence of pressure peaks and dips predicted by the model of multiple reflections, which we are going to test quantitatively in a moment (we do not have an explanation for the very first peak at $t \approx 2$ s; no peak is observed when the passenger is sitting near the train's end). Finally, at the train's exit, the pressure goes back to the constant outside value; this is the plateau at the right end (exit moment indicated by a black arrow).

From this, one can first determine the train velocity. With the time of passage through the tunnel ($t_p = 35.5$ s) and the value for the length of the tunnel (see Eq. (42.6)), one gets $v_t = 38$ m/s. This method was verified by a second independent method using the GPS sensor of the smartphone, i.e., with a method neither using the entrance and exit time stamps from the pressure curve nor the tabulated length of the tunnel, but only a value measured from the inside of the train. The method and validity of GPS-based methods of determining velocities, in turn, is well documented in the literature, including a contribution in this column [7]. Sometimes the train velocity in long distance trains (in Germany ICE) is displayed as information for the passenger. In our case we do not have this information.

We now turn to the time difference Δt between two consecutive pulses. Considering the dips in Fig. 42.1 (arrival of R pulses), one has

$$\Delta t = 7.2 \text{ s } (\pm 5\%, \text{ from the width of the minima}). \qquad (42.7)$$

Inserting this and the above values for v_s and t_p in Eq. (42.5*) (note that we do not need v_t here), one obtains

$$l = 1370 \text{ m } (\pm 5\%), \qquad (42.8)$$

which is less than 2% off (higher than) the official tabulated value, i.e., we obtain a good agreement within the error margins.[1,2] Of course, the same result is obtained by inserting v_t from Eq. (42.3) and using Eq. (42.5) instead of Eq. (42.5*). Here, we have amused ourselves to show that one can infer the unknown length of a tunnel l from measurements that can entirely be made within a moving train. If (as was actually the case here) one knows l, one can of course also calculate the theoretical value of Δt from Eq. (42.1) and compare it to the empirical values from Fig. 42.1.

---

[1] More complete error considerations can be obtained from the authors.

[2] That the experimental value lies above the official value can be understood partly because reflections of sound waves at the open end occur a bit outside (end correction). The acoustical length of the tunnel is therefore some meters greater than the physical length.

To understand the full beauty of Fig. 42.1 requires some refinements (e.g., that a rarefication wave originates from the train's rear end), and it also offers some extensions of the approach presented here (e.g., for other train types) [8].

References

1. Crawford, F.S.: Waves (Berkeley Physics Course), p. 240. McGraw Hill, New York (1968)
2. Yarmus, L.: Pulsed waves: reflections and the speed of sound. Am. J. Phys. **64**(7), 903–906 (1996)
3. Kasper, L., Vogt, P., Strohmeyer, C.: Stationary waves in tubes and the speed of sound. Phys. Teach. **53**, 52–53 (2015)
4. Hirth, M., Kuhn, J., Müller, A.: Measurement of sound velocity made easy using harmonic resonant frequencies with everyday mobile technology. Phys. Teach. **53**, 120–121 (2015)
5. https://ogy.de/androsensor
6. Muralidharan, K., Khan, A.J., Misra, A., Balan, R.K., Agarwal, S.: Barometric phone sensors: more hype than hope! In: Proceedings of the 15th Workshop on Mobile Computing Systems and Applications, p. 12. ACM, New York (2014)
7. Gabriel, P., Backhaus, U.: Kinematics with the assistance of smartphones: measuring data via GPS-visualizing data with Google Earth. Phys. Teach. **51**(4), 246–247 (2013)
8. Hirth, M.: Akustische Untersuchungen mit dem Smartphone und Tablet-Computern-Fachliche und didaktische Aspekte. Dr. Hut Publisher, Munich (2019)

Smartphone-Aided Measurements of the Speed of Sound in Different Gaseous Mixtures

43

Sara Orsola Parolin and Giovanni Pezzi

Here we describe classroom-based procedures aiming at the estimation of the speed of sound in different gas mixtures with the help of a plastic drain pipe and two iPhones or iPod touches. The procedures were conceived to be performed with simple and readily available tools.

The speed of sound in a gas, at constant temperature, is inversely proportional to the square root of the gas density [1]. The density of air, helium, and carbon dioxide, at 0 °C and standard atmospheric pressure, are respectively [2] 1.293, 0.1785, and 1.977 kg/m^3. This implies that, at constant temperature, the speed of sound is different. At 0 °C and standard atmospheric pressure, the values of the speed of sound are: 331 m/s (in air), 965 m/s (in helium), and 259 m/s (in carbon dioxide) [1].

43.1 Measurement of the Speed of Sound Wave in Different Gases

The procedure we have devised clearly shows this difference of sound speed in different gaseous mixtures and requires only two smartphones (with their applications, here-after "apps"), dry ice, helium, and PVC drain pipes.

We used the app Acoustic Ruler Pro, which is used to measure distances in the air up to 25 m [3]. In "dual device" mode, the app operates recording the delay of sound waves emitted from an iPhone, an iPod touch, or an iPad, set as "master," and reflected by another, set as a "reflector" (Fig. 43.1).

S. O. Parolin (✉)
Riolo Terme, Ravenna, Italy

G. Pezzi
Faenza, Ravenna, Italy

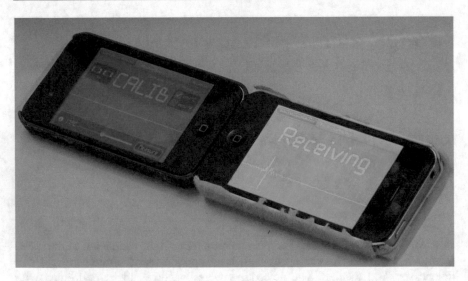

Fig. 43.1 Apparatus calibration in dual device mode: the "master" and the "reflector" devices are placed next to one another, with microphones facing each other, leaving a small gap (around 1 mm) between them, with the volume turned up on both the devices

The distance d between the two smartphones is given by the product of the speed of sound in air for half of the time interval[1]:

$$d = v\frac{\Delta t}{2}$$

Before performing measurements, the room temperature must be recorded and set.

The application always uses the speed of sound in air to calculate the distance. If smartphones are immersed in a different gas, you will get different values of the same distance. If, for example, the gas is less dense than air, sound speed will be higher. Thus, the sound wave takes less time to travel the same distance, leading to a distance result estimated by the app that is less than the actual distance. However, based on the known real distance between the two phones and the speed of sound in air, the elapsed time can be calculated as

$$\Delta t = d_{\text{estimated by iPhone}}/v_{\text{air}}$$

and used to evaluate the actual speed of sound,

$$v_{\text{effective}} = d_{\text{real}}/\Delta t.$$

---

[1] The developer of the app reports, for a quiet setting in dualdevice mode, an uncertainty on distance of ± 1 cm with a range of 25 m.

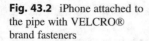

Fig. 43.2 iPhone attached to the pipe with VELCRO® brand fasteners

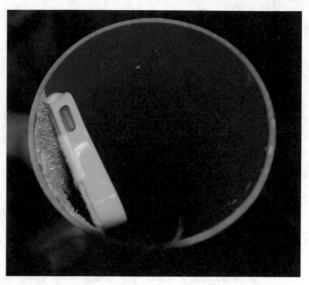

43.2 Experiment Setup

The experiments were carried out with air, with flygas (a mixture of helium and other gases used to inflate the balloons) and air, and with air and carbon dioxide. The mixtures were introduced in two stacked plastic pipes, 100 mm wide and 2 m long. The tubes also had the purpose of reducing the room noise. The enclosures with the iPhones were fixed inside the tubes by means of self-adhesive VELCRO® brand fasteners (Fig. 43.2).

Flygas was introduced while keeping the tube inclined and inserting the nozzle of the gas cylinder into the lower end of the tube, plugging the upper end in order to maintain the greatest amount of flygas inside the tube. In this way, the helium goes up through the tube and part of the air flows from the lower end. Since the opening of the cylinder valve and the flygas filling are very noisy steps (and this would interfere with the experiment), the valve was closed prior to measurements.

For measurements with carbon dioxide, dry ice was sublimated in a plastic container inserted into the drain pipe and kept vertical. Air tends to flow from the top as the air is lighter than the CO_2 gas.

The CO_2 gas temperature is lower than that of the room, and therefore a temperature gradient is created within the tube. The temperature, at a distance of 5 cm from the bottom of the tube, was -31 °C, while the temperature of solid carbon dioxide is -79 °C. To reduce some of this heat shock, we put the reflective iPhone (the one closest to dry ice) 30 cm far from the top of the ice container (Fig. 43.3).

Fig. 43.3 Assembly drawing

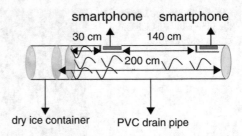

43.3 **The Measurements**

A first set of measurements was carried out when the room temperature was 17 °C.

When the distance between the two iPhones, measured with a flexible metric ruler, was 140 cm, the app distance readings for the air and flygas mixture are as reported in the second column of Table 43.1.

The speed of sound in air, upon temperature changes, can be calculated with the formula [1]

$$v = 331 \frac{m}{s} + 0.6 \frac{m}{s \cdot °C} \cdot \vartheta$$

with ϑ in °C. At the temperature measured when the experiment took place, we obtain for the speed in the air

$$v = 331 \frac{m}{s} + 0.6 \frac{m}{s \cdot °C} \cdot \vartheta = 331 \frac{m}{s} + 0.6 \frac{m}{s \cdot °C} \cdot 17 °C = 341 \frac{m}{s}.$$

It follows that, in order to cover the distance of 140 cm, the wave takes the time

$$\Delta t = \frac{\Delta s}{v} = \frac{1.40 \, m}{341 \, m/s} = 4.09 \cdot 10^{-3} \, s.$$

Table 43.1 Measurements of the speed of sound in air and flygas mixture

Measure	d (cm)	$\Delta t = d/v_{air}$ (ms)	$v = \Delta s/\Delta t$ (m/s)
1	99	2.90	483
2	81	2.37	591
3	75	2.19	639
4	70	2.05	683
5	71	2.08	673
6	70	2.05	683
7	86	2.52	556
8	83	2.43	576

The time taken by the wave to travel the actual 140-cm distance in the presence of flygas is reported, for the various measurements, in the third column of Table 43.1. The last column calculates the velocity of the gas.

The average value of the last column is 611 m/s, a result significantly lower than the speed of sound in pure helium at that temperature (979 m/s). Nonetheless, this result seems reasonable, considering that the mixture used was not composed of pure helium; in any case, the result is significantly higher than the speed of sound in air.

With the same experimental setup, a different series of measurements has been carried out using a gas mixture with carbon dioxide ($\vartheta = 20\,°C$). Average values of speed of sound between 267 and 256 m/s could be measured that are not very far away from the reference data ($v = 265$ m/s at $\vartheta = 15\,°C$).

43.4   Conclusions

Despite the experimental constraints as noted above, the developed procedure has the merit to highlight clearly, quickly, and with simple equipment that the speed of sound varies upon variation of the composition of gas mixture. A lower gas density leads to a higher speed of sound; conversely, more dense gas mixtures reduce the speed of the sound wave.

References

1. Halliday, D., Resnick, R., Walker, J.: Fundamentals of Physics, vol. 1, 8th edn. Wiley, Boca Raton, FL
2. Koshkin, N.I., Shirkevich, M.G.: Handbook of Elementary Physics, p. 45. Mir Publishers, Moscow (1977)
3. https://ogy.de/acousticruler

Part VIII

Resonators

Measurement of Sound Velocity Made Easy Using Harmonic Resonant Frequencies with Everyday Mobile Technology

44

Michael Hirth, Jochen Kuhn, and Andreas Müller

Recent contributions about smartphone experiments have described their applications as experimental tools in different physical contexts [1–4]. They have established that smartphones facilitate experimental setups, thanks to the small size and diverse functions of mobile devices, in comparison to setups with computer-based measurements. In the experiment described in this work, the experimental setup is reduced to a minimum. The objective of the experiment is to determine the speed of sound with a high degree of accuracy using everyday tools. An earlier chapter proposes a time-of-flight method where sound or acoustic pulses are reflected at the ends of an open tube (Chap. 41) [5]. In contrast, the following experiment idea is based on the harmonic resonant frequencies of such a tube, simultaneously triggered by a noise signal.

M. Hirth (✉)
Dr. Carl-Hermann-Gymnasium, Schönebeck, Germany
e-mail: hirth@gym-hermann.bildung-lsa.de

J. Kuhn
Ludwig-Maximilians-Universität München (LMU Munich), Faculty of Physics, Chair of Physics Education, Munich, Germany
e-mail: jochen.kuhn@lmu.de

A. Müller
Faculté des Sciences/Section de Physique, Institut Universitaire de Formation des Enseignants (IUFE), Université de Genève, Geneva, Switzerland
e-mail: Andreas.Mueller@unige.ch

44.1 Theoretical Background and Execution of the Experiments

Standing Waves and End Correction in a Tube

If a sound wave with a constant frequency propagates in a tube with a length L and a circular cross section with a radius R, it causes standing pressure waves in the air within the tube. The frequencies f_k of the proper modes of the tube open at both ends are the integer multiples of the fundamental frequency f_0 given by:

$$f_0 = \frac{c}{2 \cdot (L + 2a)}$$

$$f_k = (k + 1) \cdot f_0 \quad k = 0, 1, 2, \ldots$$

(44.1)

Here, c is the speed of sound in air and $a = 0.61R$ is the end correction due to the finite diameter of the tube [6]. It takes account of the fact that the pressure nodes are situated slightly outside the tube at the open end; thus, the acoustic length is longer than the geometric length of the tube [7].

In contrast, the modal frequencies of tubes only open at one end are odd-numbered multiples of another fundamental frequency and given by:

$$f_0 = \frac{c}{4 \cdot (L + a)}$$

$$f_k = (2k + 1) \cdot f_0 \quad k = 0, 1, 2, \ldots$$

(44.2)

Note that compared to Eq. (44.1) only a and not $2a$ appears in the end correction, as there is only one open end (on a closed end, no correction is necessary).

Materials and Methods for Determining the Fundamental Frequency

The cardboard tube around which aluminum foil or paper towels are usually wrapped is used for the experiments. Many other cylindrical tubes found in households are also suitable, for example, ballpoint pen tubes or gift wrap cardboard tubes. Rather than using pure tones, we generate a noise signal by rustling a piece of paper or blowing at the edge of an opening (Fig. 44.1). The continuous frequency spectrum causes all resonant frequencies to occur simultaneously. A tube with one end open is obtained by tightly sealing one end of the tube by hand.

A spectrogram (or sonogram) is used to measure the frequency spectrum (or sound signal); both can be displayed using the SpectrumView Plus app [8]. In the app, the audio sample rate is set to 16 kHz and the FFT arrangement is set to 13 (the equivalent of $2^{13} = 8192$ samples), enabling us to conduct the experiment with a frequency resolution of 16 kHz/8192 $\approx$ 2 Hz. The app Spectrum Analyzer [9] can be used if working with Android-based smartphones; however, it does not display spectrograms.

Fig. 44.1 Experimental setup: The noise caused by crumpling paper generates standing waves in a tube. The resonant frequencies are plotted by a spectrogram using a smartphone or tablet PC. Alternatively, the noise can be caused by blowing at the edge of a tube

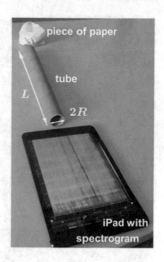

In the experiments, the respective fundamental frequency is determined using the overtones. To this end, the harmonics are visually aligned as well as possible with the equidistant lines provided by the Spectrum View Plus app. Be aware that in the app settings the button "use fixed frequency axis ticks" is switched on. This way of determining the fundamental frequency involves examining the resonant frequencies, which is easy for students to understand. However, this method can result in random measurement deviations. In our experience, the equidistant lines provided by the app can be harmonized with the measured resonant frequency bands with an accuracy of ± 10 Hz.

44.2 Experiment Analysis

Resonant Frequencies and Determining the Speed of Sound with the Tube Open at Both Ends

As can be seen in Fig. 44.2a on the left, rustling paper indeed causes a continuous spectrum. After the noise has passed through the tube, discrete frequency bands can be detected; their average frequencies correspond to the integer multiples of a fundamental frequency. For the cardboard tube with a length $L = 30.5$ cm and an opening radius $R = 1.35$ cm, the fundamental frequency is $f_0 = 533$ Hz (at a temperature of approximately 20 °C). Applying Eq. (44.1) for f_0, the speed of sound is calculated to be (343 ± 9.63) m/s (using the arithmetic law of error propagation with $\Delta f_0 = 10$ Hz, $\Delta L = \Delta R = 0.001$ m; literature value: $c = 343$ m/s at a temperature of 20 °C).

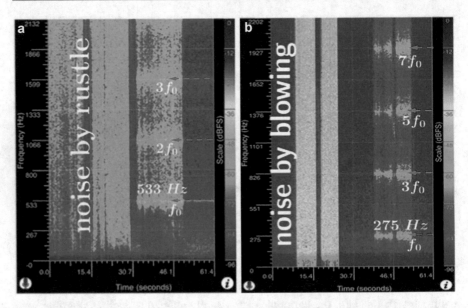

Fig. 44.2 (**a**) Left: Simply rustling a piece of paper generates a continuous spectrum. Right: Discrete frequency bands after the noise has passed through the tube. (**b**) Left: Continuous spectrum of noise produced by blowing. Right: Discrete frequency bands generated by blowing at the edge of one end of a cardboard tube that is closed at the other end

Resonant Frequencies and Determining Speed of Sound with Tube Open at One End

One end of the tube is tightly sealed with a hand, while blowing carefully at the edge of the opening at the other end. In this case, a standing wave is generated in a tube closed at one end. It can be seen in the spectrogram in Fig. 44.2b that only standing waves with odd-numbered multiples of the fundamental frequency are generated in addition to fundamental oscillation. For the same cardboard tube ($L = 30.5$ cm, radius $R = 1.35$ cm), the fundamental frequency is $f_0 = 275$ Hz (at a temperature of approximately 20 °C). This is inserted in Eq. (44.2) for f_0, yielding a value of $c = (345 \pm 14.73)$ m/s (using arithmetic law of error propagation; see above).

As an alternative or additional experiment, it is possible to examine the frequency spectrum that is plotted in the Spectrum View Plus app synchronously with the spectrogram. An example of this is shown in Fig. 44.3 for an experiment with a tube closed at one end in which the experimenter blows at the edge of the other end of the tube (Fig. 44.2b).

A detailed discussion of the possibilities of using spectrograms for acoustical analysis with mobile devices and the relevant errors is presented in Ref. [10].

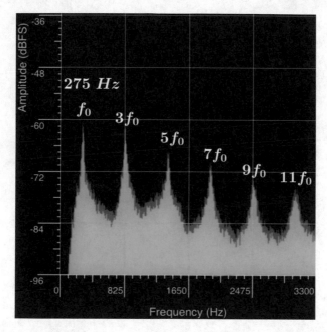

Fig. 44.3 Frequency spectrum of the sound radiated when the experimenter blows at the edge of a cardboard tube closed at one end

References

1. Klein, P., Hirth, M., Gröber, S., Kuhn, J., Müller, A.: Classical experiments revisited: Smartphone and tablet PC as experimental tools in acoustics and optics. Phys. Educ. **49**(4), 412–418 (2014)
2. Kuhn, J., Molz, A., Gröber, S., Frübis, J.: iRadioactivity–possibilities and limitations for using smartphones and tablet PCs as radioactive counters. Phys. Teach. **52**, 351–356 (2014)
3. Kuhn, J., Vogt, P., Müller, A.: Analyzing elevator oscillation with the smartphone acceleration sensors. Phys. Teach. **52**, 55–56 (2014)
4. Silva, N.: Magnetic field sensor. Phys. Teach. **50**, 372–373 (2012)
5. Kasper, L., Vogt, P., Strohmeyer, C.: Stationary waves in tubes and the speed of sound. Phys. Teach. **53**, 52–53 (2015)
6. Levine, H., Schwinger, J.: On the radiation of sound from an unflanged circular pipe. Phys. Rev. **73**(4), 383–406 (1948)
7. Vogt, P., Kasper, L., Fahsl, C., Herm, M., Quarthal, D.: Physics2Go! Den Alltag mit dem Smartphone entdecken. In: Bresges, A., Mähler, L., Pallack, A. (eds.) Themenspezial MINT, pp. 46–60. MNU, Publisher Klaus Seeberger, Neuss, Germany (2015)
8. https://ogy.de/SpektrumView
9. https://ogy.de/SpektrumAnalyzer
10. Hirth, M., Gröber, S., Kuhn, J., Müller, A.: Experimental investigation of acoustical resonances in one-dimensional shaft carriers with Smartphone and Tablet-PC. Physik. **1**(14), 12–25 (2015)

Corkscrewing and Speed of Sound: A Surprisingly Simple Experiment

45

Lutz Kasper and Patrik Vogt

The first purpose of uncorking a bottle of wine probably is to have good prospect of culinary pleasure. But if you are a physicist or physics teacher you may have a secondary interest. Couldn't it be possible to determine the speed of sound just by uncorking the bottle and with a little help from a smartphone? Sure, it is. Here comes the idea.

45.1 Determining the Speed of Sound: A Five-Second Smartphone Experiment

Most people know well the audible plop noise while opening a wine bottle. So, how is this 'plop' actually caused and on what does it depend? Lastly, what information can we obtain from this noise?

The process of uncorking a bottle is characterized by friction between the cork and the inner side of the bottleneck. Furthermore, there is an airflow due to a change of gas pressure. Those effects generate a multi-frequency noise. From an acoustical point of view, the bottleneck with the wine level is a one-side-closed pipe (see Fig. 45.1).

If we consider the bottleneck a resonator like in Fig. 45.1, then we should expect resonance frequencies. Although we can use freeware computer apps such as Audacity [1] to measure the frequency spectrum, smartphone apps offer advantages in terms of mobility (Chap. 41) [2, 3]. Appropriate apps for example are Spektroskop (iOS) [4] and phyphox (Android and iOS) [5].

L. Kasper (✉)
University of Education Schwäbisch Gmünd, Schwäbisch Gmünd, Germany
e-mail: lutz.kasper@ph-gmuend.de

P. Vogt
Institute of Teacher Training (ILF) Mainz, Mainz, Germany
e-mail: vogt@ilf.bildung-rp.de

Fig. 45.1 The bottleneck as a
one-side-closed pipe resonator

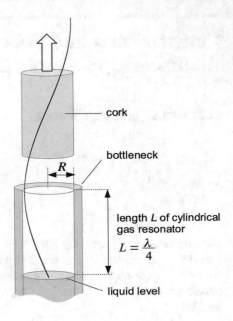

cork

bottleneck

R

length L of cylindrical
gas resonator

$$L = \frac{\lambda}{4}$$

liquid level

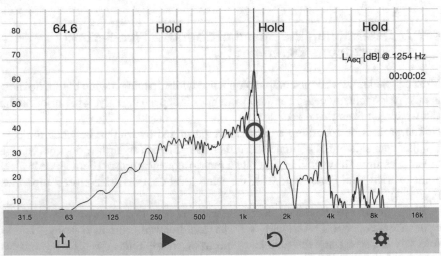

Fig. 45.2 Resonance frequency when uncorking a wine bottle (app: Spektroskop)

As expected the frequency measurement while uncorking the bottle shows a clear peak in the FFT diagram (Fig. 45.2). There are oscillating nodes at the closed end of one-side-closed pipes and antinodes at the open end (Fig. 45.1). For the fundamental frequency f_0, the length L of the resonator corresponds to $\lambda/4$ (λ: wavelength). From this we get

$$f_0 = \frac{1}{4L} c_{\text{gas}}$$

(L: length of resonator, c_{gas}: speed of sound in the gas).

As a simplification, it is assumed that the residual gas in the bottleneck is equivalent to air. Although a more exact examination would show an amount of ethanol as well, an experimental comparison with water-filled bottles can justify this simplification. Therefore, in the following we will use c_{air} instead of c_{gas}.

However, in another respect we have to include a correction. For more accurate results, the end-pipe correction ΔL needs to be taken into account (Chap. 41) [6, 7]. ΔL depends on radius R of the bottleneck:

$$\Delta L = 0.61R$$

Therefore, the speed of sound is given by

$$c_{\text{air}} = 4 f_0 (L + \Delta L)$$

The measurement in Fig. 45.2 was performed on a wine bottle with a resonator length of 6 cm and an inner diameter of 2 cm ($R = 0.010$ m). By inserting all parameters in the equation above, we obtain the speed of sound $c_{\text{air}} = 332$ m/s. Since this experiment was carried out at a room temperature of 23 °C, we should expect a value for the speed of sound $c = 344.8$ m/s [8]. So, the average relative error is under 5%, which is pretty acceptable for this five-second freehand experiment.

45.2   The Experiment for Use in Physics Classroom

Opening wine bottles in a classroom? Perhaps this is not a good idea. Once you have uncorked a bottle at home—for scientific reasons—you can use the empty bottle for this experiment too. You can even expand the possibilities by adding cm marks at the bottleneck (Fig. 45.3). Now students systematically can vary the resonator length by filling in water with different levels. Instead of corkscrewing, one can use a wet finger to generate the plop noise. Moreover, in this version there is no alcohol vapor anymore and no simplification is needed.

Measurement samples for a water-filled bottle with different resonator lengths are given in Table 45.1.

The room temperature for this experiment was 24 °C. The theoretically expected value for the speed of sound in air should be $c = 345.6$ m/s [8]. Here again, the average relative error is roughly 5%, which is not bad for this kind of simple experiment.

Fig. 45.3 Prepared
bottleneck for
experimentation

Table 45.1 Fundamental
frequency and speed of
sound as a function of resonator length
($R = 0.010$ m)

L (m)	f_0 (measured) (Hz)	c_{air} (m/s)
0.030	2355	340
0.040	1740	321
0.050	1488	334
0.060	1225	324
0.070	1058	322
0.080	952	328

References

1. Audacity (freeware). http://www.audacityteam.org (Windows, MacOS, Linux).
2. Hirth, M., Kuhn, J., Müller, A.: Measurement of sound velocity made easy using harmonic resonant frequencies with everyday mobile technology. Phys. Teach. **53**, 120–121 (2015)
3. Kuhn, J., Vogt, P.: Smartphone & Co. in physics education: effects of learning with new media experimental tools in acoustics. In: Schnotz, W., Kauertz, A., Ludwig, H., Müller, A., Pretsch, J. (eds.) Multidisciplinary Research on Teaching and Learning, pp. 253–269. Palgrave Macmillan, Basingstoke, UK (2015)
4. https://ogy.de/Spektroskop
5. https://phyphox.org/de/download-de/
6. Levine, H., Schwinger, J.: On the radiation of sound from an unflanged circular pipe. Phys. Rev. **73**, 383 (1948)
7. Kasper, L., Vogt, P., Strohmeier, C.: Stationary waves in tubes and the speed of sound. Phys. Teach. **53**, 523–524 (2015)
8. Temperature dependency of sound velocity in air. https://ogy.de/engineeringtoolbox

A Bottle of Tea as a Universal Helmholtz Resonator

46

Martín Monteiro, Cecilia Stari, Cecilia Cabeza, and Arturo C. Marti

Resonance is an ubiquitous phenomenon present in many systems. In particular, air resonance in cavities was studied by Hermann von Helmholtz in the 1850s. Originally used as acoustic filters, Helmholtz resonators are rigid-wall cavities that reverberate at given fixed frequencies. An adjustable type of resonator is the so-called universal Helmholtz resonator, a device consisting of two sliding cylinders capable of producing sounds over a continuous range of frequencies. Here we propose a simple experiment using a smartphone and normal bottle of tea with a nearly uniform cylindrical section, which, when filled with water at different levels, mimics a universal Helmholtz resonator. Blowing over the bottle, we notice different sounds are produced. Taking advantage of the great processing capacity of smartphones, sound spectra together with frequencies of resonance are obtained in real time.

46.1 Helmholtz Resonator

Helmholtz resonators consist of rigid-wall containers, usually made of glass or metal, with volume V and neck with diameter D and length L as indicated in Fig. 47.1. In the past, they were used as acoustic filters, for the reason that when someone blows over the opening, air inside the cavity resonates at a frequency given by

M. Monteiro (✉)
Universidad ORT Uruguay, Montevideo, Uruguay
e-mail: monteiro@ort.edu.uy

C. Stari · C. Cabeza · A. C. Marti
Universidad de la República, Montevideo, Uruguay
e-mail: cstari@fing.edu.uy; cecilia@fisica.edu.uy; marti@fisica.edu.uy

Fig. 47.1 The tea bottle, with
a nearly uniform circular
section, is suitable for the
present experience and
dimensions: $D = 29.0(2)$ mm
and $L = 70(1)$ mm

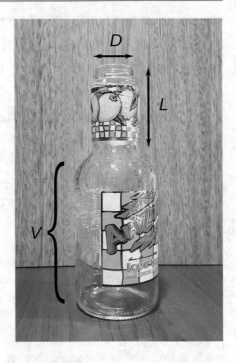

$$f = \frac{c}{2\pi} \sqrt{\frac{A}{VL'}},$$

where $A = \pi \cdot (D/2)^2$ is the section of the neck. In this expression c is the sound speed
and L' is the equivalent length of the neck, accounting for the end correction, which,
in the case of outer end unflanged, results in $L' = L + 1.46a$ (with $a = D/2$). Please
see the final note about end correction. The previous expressions are valid provided
that the linear dimensions are smaller than the typical wavelength ($L \ll \lambda$, $a \ll \lambda$,
$h \ll \lambda$) (h is the height of the cavity). A universal Helmholtz resonator is a special
type of resonator in which the volume can be varied, resulting in a range of possible
frequencies of resonance.

46.2   A Smartphone-Based Experiment in Acoustics

There are several examples of physics experiments using smartphones in the field of
acoustics (e.g. Chaps. 41, 43, 44, 47 and 48) [1–7]. In some of them the smartphone
is used as a microphone to digitalize sound or as a source of pure or more complex
sounds (e.g. Chaps. 41 and 48) [1, 2]. An important advantage of smartphones, not
always taken into account, is their capacity to speedily process a lot of information
and, in particular, to obtain sound spectra in real time (Chaps. 43, 44 and 47) [3–7].

To obtain the volume of the cavity, the experiments starts by placing the empty
bottle on a weighing scale and setting it to zero. After that, the bottle is filled with

Fig. 47.2 Experimental
setup: the bottle on the scale
and the smartphone displaying
a sound spectrum

water until the neck (7.0 cm below the opening) and the reading on the scale, in this case 435 g, indicates the cavity volume, that is, $V = 435$ cm^3.

Next, the bottle is filled with water at different levels. For each water level, the volume of the air cavity is obtained by subtracting the added water to the total volume of the cavity. Blowing across the top of the bottle, with the lower lip touching the edge of the bottle, produces a resonance in the cavity. The Advanced Spectrum Analyzer PRO app is used to obtain the spectrum in real time and the highest peak that corresponds to the resonance frequency, as shown in Fig. 47.2. The "hold" button is useful to freeze the display and write down the frequency. So, varying the volume of water in the bottle, it is possible to find the relationship between volume and resonance frequency.

46.3   Results and Analysis

In Fig. 47.3 we show the linearized relationship between the resonance frequency and the cavity volume; specifically, we plot the f^2 vs. $1/V$. Note that corresponding wavelengths, between 1 and 1.5 m, are much larger than the linear dimensions of the cavity.

The sound speed c is related to the slope according to

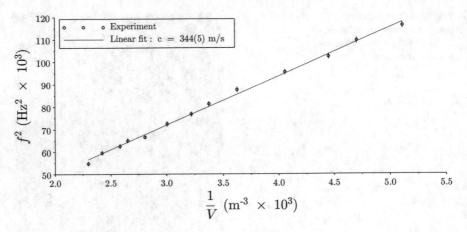

Fig. 47.3 Linearized relationship between resonance frequency and air volume: experimental values and linear fit

$$\text{slope} = \frac{c^2 A}{4\pi^2 (L + 1.46a)}.$$

In our experiment, the sound speed obtained is $c = 344(5)$ m/s, resulting in an uncertainty of less than 2%. Taking into account that the ambient temperature was 22.3 °C, the value exhibits a great coherence with reference value, 344 m/s, at this temperature. We note that there are at least a couple of uncontrollable sources of uncertainties in this experience: the temperature of the blown air is a bit larger than the ambient temperature and the contour conditions at the opening are modified by the blow and the proximity of the lips.

To sum up, we presented a very simple experiment using everyday stuff to study acoustic resonance. This experiment is based on the great processing capacity of smartphones to obtain real-time spectra. The result obtained for the sound speed is in concordance with the standard value.

46.4 Note About the End Correction

The inner end of the neck is flanged, so the correction factor is $8/(3\pi)$ (according to Kinsler, Ref. [8]). While the outer end of the neck is unflanged, then the correction factor is 0.6133 (according to Levine and Schwinger, Ref. [9]). Thus, the total correction factor for the case of the bottle is the sum of these two factors, that is 1.4621.

References

1. Kuhn, J., Vogt, P.: Analyzing acoustic phenomena with a smartphone microphone. Phys. Teach. **51**, 118–119 (2013)
2. Yavuz, A.: Measuring the speed of sound in air using smartphone applications. Phys. Educ. **50**(3), 281 (2015)
3. Kasper, L., Vogt, P., Strohmeyer, C.: Stationary waves in tubes and the speed of sound. Phys. Teach. **53**, 52–53 (2015)
4. Parolin, S.O., Pezzi, G.: Smartphone-aided measurements of the speed of sound in different gaseous mixtures. Phys. Teach. **51**, 508–509 (2013)
5. Monteiro, M., Martí, A.C., Vogt, P., Kasper, L., Quarthal, D.: Measuring the acoustic response of Helmholtz resonators. Phys. Teach. **53**, 247–249 (2015)
6. Hirth, M., Kuhn, J., Müller, A.: Measurement of sound velocity made easy using harmonic resonant frequencies with everyday mobile technology. Phys. Teach. **53**, 120–121 (2015)
7. González, M., González, M.: Smartphones as experimental tools to measure acoustical and mechanical properties of vibrating rods. Eur. J. Phys. **37**(4), 045701 (2016)
8. Kinsler, L.E., Frey, A.R., Coppens, A.B., Sanders, J.V.: Fundamentals of Acoustics, 4th edn. Wiley-VCH (1999)
9. Levine, H., Schwinger, J.: On the radiation of sound from an unflanged circular pipe. Phys. Rev. **73**(4) (1948)

Measuring the Acoustic Response of Helmholtz Resonators

47

Martín Monteiro, Arturo C. Marti, Patrik Vogt, Lutz Kasper, and Dominik Quarthal

Many experiments have been proposed to investigate acoustic phenomena in college and early undergraduate levels, in particular the speed of sound [1–9], by means of different methods, such as time of flight, transit time (Chap. 40), or resonance in tubes (Chap. 41). In this chapter we propose to measure the acoustic response curves of a glass beaker filled with different gases, used as an acoustic resonator. We show that these curves expose many interesting peaks and features, one of which matches the resonance peak predicted for a Helmholtz resonator fairly well, and gives a decent estimate for the speed of sound in some cases. The measures are obtained thanks to the capabilities of smartphones.

M. Monteiro (✉)
Universidad ORT Uruguay, Montevideo, Uruguay
e-mail: monteiro@ort.edu.uy

A. C. Marti
Universidad de la República, Montevideo, Uruguay
e-mail: marti@fisica.edu.uy

P. Vogt
Institute of Teacher Training (ILF) Mainz, Mainz, Germany
e-mail: vogt@ilf.bildung-rp.de

L. Kasper
Department of Physics, University of Education Schwäbisch Gmünd, Schwäbisch Gmünd, Germany
e-mail: lutz.kasper@ph-gmuend.de

D. Quarthal
Department of Chemistry, University of Education Freiburg, Freiburg, Germany
e-mail: dominik.quarthal@ph-freiburg.de

© The Author(s), under exclusive license to Springer Nature Switzerland AG 2022
J. Kuhn, P. Vogt (eds.), *Smartphones as Mobile Minilabs in Physics*,
https://doi.org/10.1007/978-3-030-94044-7_47

47.1   Execution of the Experiment

For the experiment we used a typical beaker found in almost all science laboratories, as shown in Fig. 47.1. The beaker has height $H = 0.139$ m and radius $a = 0.040$ m. A similar setup is attainable with simple glasses at home (Fig. 47.2). The beaker was filled with the gas to be measured, side up if the gas is denser than air or side down if the gas is lighter than air.

Two smartphones were employed in this experiment. One smartphone near the beaker was used as a white-noise generator to stimulate the resonant modes in the cavity (black phone just visible to the left behind the white phone in Fig. 47.2). The other smartphone was used to record the sound (white phone in Fig. 47.2). Its location is a delicate matter; it must be placed close to the beaker to obtain a good signal but care should be taken to not modify the boundary conditions. We obtained better results when the smartphone protruded slightly into the glass (Fig. 47.2). This will change (although minimally) the volume and opening area, but the spectrum is measured with much higher accuracy. The measures of the acoustic response must be made with apps that perform a fast Fourier transform in real time. For this purpose we used the app Spektroskop [10] on an iPhone. On Android phones the same can be done with apps like Advanced Spectrum Analyzer [11] that automatically detect the peak frequencies.

The acoustic spectral response of the beaker was obtained for air and three other different gases: oxygen, carbon dioxide, and methane (Figs. 47.3, 47.4, 47.5, and 47.6), showing subtle and interesting spectral differences between the four gasses. Different features can be appreciated by means of a little theoretical analysis.

First, the acoustic response of the beaker can be understood by thinking of the beaker as a resonant cavity. The calculus of the normal modes of resonance is

Fig. 47.1 In physics labs with different gases (left: heavier than air; right: lighter than air)

Fig. 47.2 Investigation with
glasses at home

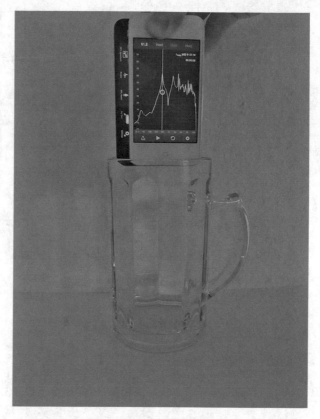

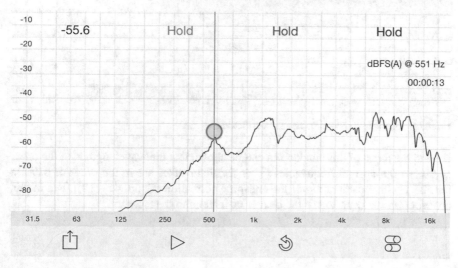

Fig. 47.3 Frequency spectrum with air

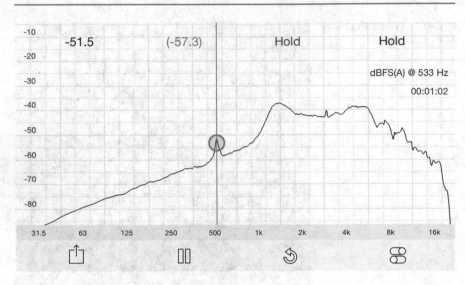

Fig. 47.4 Frequency spectrum with oxygen

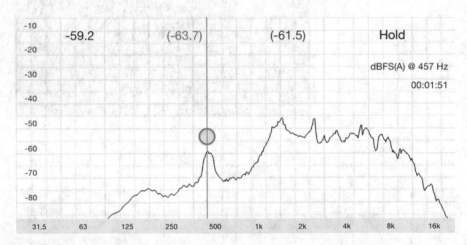

Fig. 47.5 Frequency spectrum with carbon dioxide

beyond the focus of this chapter but can be shown that the wavelength is of the order of $4H$, so the normal frequencies in air for this beaker are in the order of 617 Hz [or 525 Hz in consideration of the end-pipe correction (Chap. 41) [12]]. But in this contribution we want to see the beaker as a Helmholtz resonator.

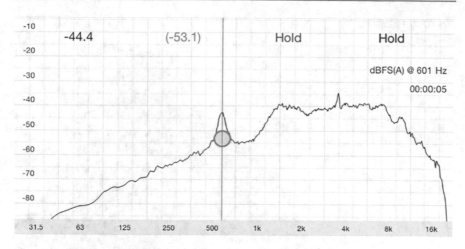

Fig. 47.6 Frequency spectrum with methane

47.2   Theoretical Background and Experiment Analysis

A rigid cavity with an open neck can be modeled as a mass-spring system, where the cavity is the spring and the neck is the mass, the so-called "Helmholtz resonator" (Fig. 47.7). The only frequency of this system is given by [13, 14].

$$f = \frac{c}{2\pi} \sqrt{\frac{A}{V \cdot L'}},\tag{47.1}$$

where L' is the effective length of the neck, A is the area of the neck, V is the volume of the cavity, and c is the speed of sound in the inner gas.

Because a little amount of mass of gas is moving outside the edges of the neck dragged by the gas inside the neck, the effective length of the neck L' is slightly greater than the physical length of the neck L. This end correction depends on the boundary conditions [12–16]

$$\text{with an outer end flanged} : L' = L + 1.7a,\tag{47.2a}$$

$$\text{with an outer end unflanged} : L' = L + 1.4a,\tag{47.2b}$$

where a is the radius of the opening. In our case, the resonator is a cylindrical glass, then the neck has a real null length $L = 0$; thus, the effective length must be expressed completely by the end correction of a flanged border. Moreover, by

Fig. 47.7 Helmholtz
resonator

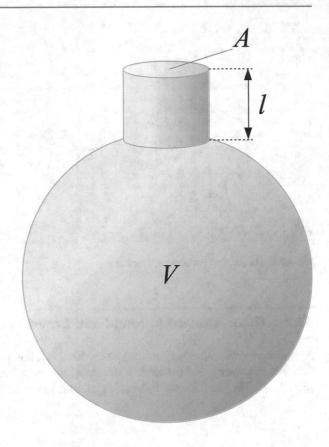

means of this particular geometry, the volume of the cavity is $V = AH$, where H is the
height of the cylinder. Then, with all that in mind, the resonant frequency is

$$f = \frac{c}{2\pi}\sqrt{\frac{1}{1.7aH}}. \tag{47.3}$$

We then assume that the best defined peak in the middle region of the spectrum
corresponds to the Helmholtz frequency (Eq. 47.3). So, from this peak, we can
determine the speed of sound, which can be expressed as a function of this frequency
of resonance:

$$c = 2\pi f\sqrt{1.7aH}. \tag{47.4}$$

Let us call this the measured speed of sound.

In order to compare, remember that assuming that the gas is an ideal gas, the
speed of sound is given by the well-known thermodynamical relation [17]

Table 47.1 Experimental results for a temperature of 17.5 °C

Gas	Measured resonance frequency (Hz)[a]	Speed of sound: measured (m/s)	Speed of sound: reference values (m/s)	Deviation (%)
Air ($\gamma = 1.4$, $M = 0.029$ kg/mol)	557 ± 3	340 ± 2	341	0.3
Oxygen ($\gamma = 1.4$, $M = 0.032$ kg/mol)	533 ± 3	325 ± 2	325	0.0
Carbon dioxide ($\gamma = 1.4$, $M = 0.044$ kg/mol)	457 ± 3	279 ± 2	277	0.7
Methane ($\gamma = 1.3$, $M = 0.016$ kg/mol)	598 ± 3	365 ± 2	443	17.6

[a]The frequencies were not read from the display, but from the exported data

$$c = \sqrt{\frac{\gamma RT}{M}}, \tag{47.5}$$

where γ is the adiabatic index of the gas, R is the universal gas constant (8.31 J·mol^{-1}·K^{-1}), T is the absolute temperature in kelvin, and M is the molar mass of the gas.

The resonance frequencies of the best defined peaks in the middle region of the spectra and the calculated results from Eqs. (47.4) and (47.5) are shown in Table 47.1 for the aforementioned gases. As can be seen, if we assume that the best defined peak in the middle frequency regime is centered at the frequency predicted by the theoretical model for an ideal Helmholtz resonator, then we get reasonable estimates for the speed of sound in each gas, with the methane case showing the largest discrepancy with the accepted values.

An interesting extension could consist of employing other gases such as helium or sulfur hexafluoride, with densities and speed of sound considerably different from those of air.

References

1. Carvalho, C.C., Lopes dos Santos, J.M.B., Marques, M.B.: A time-of-flight method to measure the speed of sound using a stereo sound card. Phys. Teach. **46**, 428 (2008)
2. Vogt, P., Kuhn, J.: Determining the speed of sound with stereo headphones. Phys. Teach. **50**, 308 (2012)
3. Bacon, M.E.: Speed of sound versus temperature using PVC pipes open at both ends. Phys. Teach. **50**, 351 (2012)
4. Bin, M.: Measuring the speed of sound using only a computer. Phys. Teach. **51**, 295 (2013)

5. Parolin, S.O., Pezzi, G.: Smartphone-aided measurements of the speed of sound in different gaseous mixtures. Phys. Teach. **51**, 508 (2013)
6. Gómez-Tejedor, J.A., Castro-Palacio, J.C., Monsoriu, J.A.: Direct measurement of the speed of sound using a microphone and a speaker. Phys. Educ. **49**, 310 (2014)
7. Aljalal, A.: Time of flight measurement of speed of sound in air with a computer sound card. Eur. J. Phys. **35**, 065008 (2014)
8. Kasper, L., Vogt, P., Strohmeyer, C.: Stationary waves in tubes and the speed of sound. Phys. Teach. **53**, 52 (2015)
9. Hirth, M., Kuhn, J., Müller, A.: Measurement of sound velocity made easy using harmonic resonant frequencies with everyday mobile technology. Phys. Teach. **53**, 120 (2015)
10. https://ogy.de/spektroskop
11. https://ogy.de/advancedspectrumanalyzerpro
12. Levine, H., Schwinger, J.: On the radiation of sound from an unflanged circular pipe. Phys. Rev. **73**, 383 (1948)
13. Kinsler, L.E., et al.: Fundamentals of Acoustics, 4th edn. Wiley, New York (2000)
14. Lüders, K., von Oppen, G.: Mechanik, Akustik, Wärme. Walter de Gruyter, Berlin (2008)
15. Ruiz, M.J.: Boomwhackers and end-pipe corrections. Phys. Teach. **52**, 73–75 (2014)
16. Fletcher, N.H., Rossing, T.D.: The Physics of Musical Instruments. Springer, New York (1998)
17. French, P.: Vibrations and Waves, p. 212. W.W. Norton, New York (1969)

Part IX

Other Acoustic Phenomena

Analyzing Acoustic Phenomena with a Smartphone Microphone

48

Jochen Kuhn and Patrik Vogt

This contribution describes how different sound types can be explored using the microphone of a smartphone and a suitable app. Vibrating bodies, such as strings, membranes, or bars, generate air pressure fluctuations in their immediate vicinity, which propagate through the room in the form of sound waves. Depending on the triggering mechanism, it is possible to differentiate between four types of sound waves: tone, sound, noise, and bang. In everyday language, non-experts use the terms "tone" and "sound" synonymously; however, from a physics perspective there are very clear differences between the two terms. This chapter presents experiments that enable learners to explore and understand these differences. Tuning forks and musical instruments (e.g., recorders and guitars) can be used as equipment for the experiments. The data are captured using a smartphone equipped with the appropriate app (in this work we describe the app Audio Kit for iOS systems [1]). The values captured by the smartphone are displayed in a screen shot and then viewed directly on the smartphone or exported to a computer graphics program for printing.

48.1 Capture and Analysis of Different Types of Sound Waves

Each sound source is examined separately, and an oscillogram and a frequency spectrum of its acoustic signals are displayed. This is produced by hitting a tuning fork with a striker, plucking a stringed instrument, blowing into a wind instrument, or scrunching up a piece of paper. Additionally, the acoustic signal generated by a

J. Kuhn (✉)
Ludwig-Maximilians-Universität München (LMU Munich), Faculty of Physics, Chair of Physics Education, Munich, Germany
e-mail: jochen.kuhn@lmu.de

P. Vogt
Institute of Teacher Training (ILF) Mainz, Mainz, Germany
e-mail: vogt@ilf.bildung-rp.de

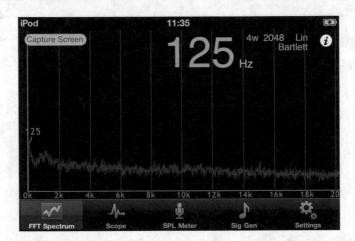

Fig. 48.1 Functions of the Audio Kit software

hand clap can also be analyzed. Recording is performed by switching on the iPhone or iPod touch and opening the appropriate software. The following is a description of the experiment using an iPod touch equipped with the Audio Kit app. Audio Kit has a menu at the bottom of the screen (Fig. 48.1) offering a selection of options, so that the user can select the desired form of representation of acoustic wave analysis, e.g., "FFT Spectrum" to display a frequency spectrum or "Scope" to display an oscillogram.

48.2 Analysis of a Tone

The acoustic signal of a tone can be determined by a sine function, which is represented in an oscillogram:

$$y = \hat{y} \cdot \sin\left(2\pi ft\right),$$

with y = displacement, $\hat{y}$ = amplitude, f = frequency, and t = time. When, for example, a tuning fork is hit with a striker, the display set to "FFT Spectrum" shows a frequency spectrum with one single spectral line (Fig. 48.2), whereas if it is set to "Scope," it displays a sinusoidal curve. If, for example, the objective is to determine the frequency of the tone, a screen shot has to be made of the sinusoidal curve. In order to determine the frequency, the vibration can either be analyzed directly using the screen shot on the smartphone or by exporting the data to a computer graphics program for printing.

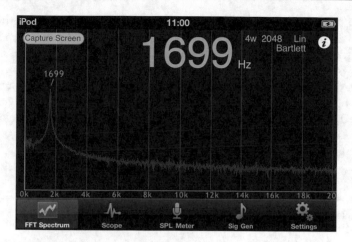

Fig. 48.2 Frequency of the tone of a 1700-Hz tuning fork

48.3   Analysis of a Sound

A musical instrument produces a periodic and not a sinusoidal vibration pattern
(Fig. 48.3); this is called a "sound." A signal of this kind results from a sum of sine
functions that are the integer multiples of a fundamental frequency f_1 (principle of
Fourier):

$$y = \sum_n \widehat{y}_n \cdot \sin\left(2\pi f_n t\right), \text{ with } f_n = nf_1 (n = 1, 2, 3, \ldots).$$

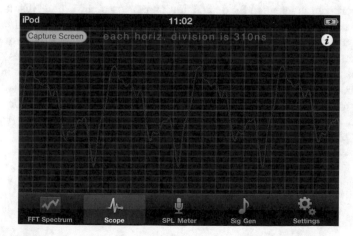

Fig. 48.3 Oscillogram of the sound of a melodica

A sound is therefore a mix of individual tones whose frequencies are integrated multiples of the lowest frequency present. The frequency spectrum also includes harmonics alongside the fundamental tone whose frequencies correspond to integer multiples of the frequency of the fundamental tone.

If a musical instrument (e.g., a recorder) is played and the sound is, for example, captured by an iPod touch and Audio Kit, the display set to "FFT Spectrum" will display a frequency spectrum of a fundamental tone as well as harmonics whose frequencies correspond to integer multiples of the frequency of the fundamental tone.

48.4 Analysis of Noise and Impulse

When paper is scrunched up, it makes a noise. Contrary to tone and sound, noise is not produced by periodic processes. The Fourier analysis conducted by an iPod touch with Audio Kit produces a continuous spectrum (rustle) when set to "FFT Spectrum," which does not distinguish between individual frequencies. Furthermore, it is not possible to identify regularity on the oscillogram ("Scope" setting).

A bang is a sudden mechanical vibration with high amplitude and a short duration. An example of this is a hand clap. The frequency spectrum of a bang also does not distinguish between individual frequencies; therefore, the frequency spectrum produced by Audio Kit set to "FFT Spectrum" displays a similar continuous spectrum to noise.

48.5 Further Information

If the experiment is intended for demonstration purposes, the iPhone display can be projected using a webcam. In addition to the experiments described in this chapter the phenomenon of acoustic color can also be explored using a variety of musical instruments. Acoustic color is influenced by both transient processes and the harmonics spectrum. Screen shots can be made, printed out, and stuck in pupils' exercise books in order to keep a record of the results.

Research shows that studying acoustic phenomena with mobile devices integrated in a more sophisticated instructional setting and combined with more everyday phenomena [2–4] could also increase learning [5].

References

1. itunes.apple.com/de/app/audio-kit/id376965050
2. Vogt, P., Kuhn, J., Neuschwander, D.: Determining ball velocities with smartphones. Phys. Teach. **52**(6), 376–377 (2014)
3. Schwarz, O., Vogt, P., Kuhn, J.: Acoustic measurements of bouncing balls and the determination of gravitational acceleration. Phys. Teach. **51**(5), 312–313 (2013)
4. Müller, A., Vogt, P., Kuhn, J., Müller, M.: Cracking knuckles – a smartphone inquiry on bioacoustics. Phys. Teach. **53**(5), 307–308 (2015)
5. Kuhn, J., Vogt, P.: Smartphone & Co. in physics education: effects of learning with new media experimental tools in acoustics. In: Schnotz, W., Kauertz, A., Ludwig, H., Müller, A., Pretsch, J. (eds.) Multidisciplinary Research on Teaching and Learning, pp. 253–269. Palgrave Macmillan, Basingstoke, UK (2015)

Analyzing the Acoustic Beat with Mobile Devices

49

Jochen Kuhn, Patrik Vogt, and Michael Hirth

Various examples of how physical relationships can be examined by analyzing acoustic signals using smartphones or tablet PCs are presented in this book (Chaps. 12, 43 and 48) [1–3] and have been proven to be helpful for learning such phenomena [4]. In this example, we will be exploring the acoustic phenomenon of beats, which is produced by the overlapping of two tones with a low difference in frequency Δf. The resulting auditory sensation is a tone with a volume that varies periodically. Acoustic beats can be perceived repeatedly in day-to-day life and have some interesting applications. For example, string instruments are still tuned with the help of an acoustic beat, even with modern technology. If a reference tone (e.g., 440 Hz) and, for example, a slightly out-of-tune violin string produce a tone simultaneously, a beat can be perceived. The more similar the frequencies, the longer the duration of the beat. In the extreme case, when the frequencies are identical, a beat no longer arises. The string is therefore correctly tuned. Using the Oscilloscope app [5], it is possible to capture and save acoustic signals of this kind and determine the beat frequency f_S of the signal, which represents the absolute value of the difference in frequency $|\Delta f|$ of the two overlapping tones.

J. Kuhn (✉)
Ludwig-Maximilians-Universität München (LMU Munich), Faculty of Physics, Chair of Physics Education, Munich, Germany
e-mail: jochen.kuhn@lmu.de

P. Vogt
Institute of Teacher Training (ILF) Mainz, Mainz, Germany
e-mail: vogt@ilf.bildung-rp.de

M. Hirth
Dr. Carl-Hermann-Gymnasium, Schönebeck, Germany
e-mail: hirth@gym-hermann.bildung-lsa.de

49.1   Theoretical Background and Execution of the Experiment

If two sine tones with very close frequencies are played simultaneously, beats are produced. On the one hand, these beats are easy to produce and their quality can be easily perceived. On the other hand, oscillograms of beats can be recorded and the relationship between the output frequency and beat frequency can be analyzed and quantified.

The acoustic beat is an example of the principle of unimpeded superposition of sound waves (the principle of superposition). Figure 49.1 illustrates this property for sine waves with frequencies that are almost equal. At (a) the oscillation amplitudes add up to a maximum, while at (b) the opposing oscillation amplitudes add up to 0.

The beat frequency of two sinusoids f_1 and f_2 can be easily deduced using a vector diagram, assuming that $f_1 < f_2$ and $2f_1 \geq f_2$ hold. At time $t_0 = 0$, both of the oscillations have the same phase position and the pointers therefore point in the same direction. After a period of beat T_S, the faster pointer with the frequency of f_2 just catches up with the slower one with the frequency of f_1 for the very first time. As a result, the following applies to the phase angle:

Fig. 49.1 Superposition of overlapping sound oscillations

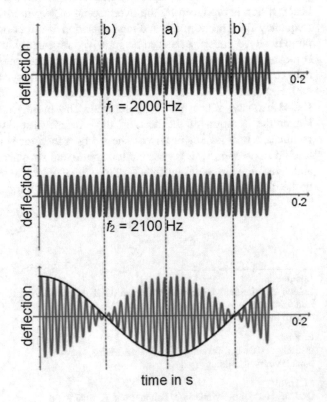

Fig. 49.2 Experimental
setup; the apps used were
AudioKit and Oscilloscope

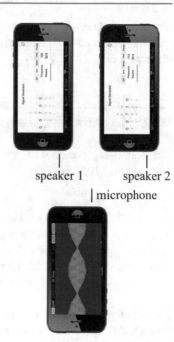

speaker 1   speaker 2

| microphone

$$\varphi_2(T_S) = \varphi_1(T_S) + 2\pi \Leftrightarrow 2\pi f_1 T_S = 2\pi f_1 T_S + 2\pi$$

With $T_S = 1/f_S$, $f_S = |f_2 - f_1|$ follows.

The perceived tone frequency f_W is determined by:

$$f_W = \frac{f_1 + f_2}{2}.$$

To analyze and quantify the physical relationships of an acoustic beat, two defined
sinusoidal acoustic signals need to be produced (and also reproduced for repeated
measures) and then analyzed. In an experiment, three smartphones are therefore
required to analyze the acoustic beat: two of them produce the sine tones with
slightly different frequencies and the third device detects and analyzes the
overlapping oscillation. To produce a beat that is as precise as possible, the
amplitudes of sound pressure of both tones need to be identical at the location
where the sound is detected. For this purpose, the speakers of both smartphones
are positioned at the same distance to the microphone of the sound-detecting device
(Fig. 49.2).

The Audio Kit app [6] sound generator is, for example, suitable for producing the
tones (for Android, e.g., Frequency Generator [7]). This app is opened on two
iPhones or iPod Touches and the volume is set to a maximum both on the app and
the devices. To record the acoustic beat produced by the superposition, a third device
is used with, for instance, the Oscilloscope app. Both of the sound generators
initially play random frequencies and you can perceive the harmony of the tones.

Then the frequencies are gradually brought closer together until they reach the extreme case and are identical. Oscillograms can be recorded and analyzed for different beats.

49.2   Experiment Analysis

The following oscillogram was produced with sine tones with frequencies of 1000 Hz and 1050 Hz (Fig. 49.3).

Qualitative When you listen to the harmony of the tones produced with $f_1 = 1000$ Hz and $f_2 = 1050$ Hz, you can perceive a "vibrating" tone. As the frequencies of the tones become more similar, you hear how the volume of the tone goes up or down periodically and how the tone itself becomes lower. The more similar the frequencies, the longer the duration of a period. If both frequencies are identical, a beat is no longer perceived.

Quantitative By recording the oscillogram of the beat with the suggested frequencies, we can link the auditory sensation (of "vibrating," periodically loud and soft) to the form of the varying amplitude. In addition, the period of the beat can be determined to be 20 ms, which is the equivalent of a beat frequency of 50 Hz. The perceived tone frequency f_W is analyzed by zooming ("pinch zoom") into the antinode and examining the oscillation period captured there. In this case the result is 0.98 ms, which corresponds to a perceptible tone frequency f_W of approximately 1020 Hz, showing more deviation from the theoretical value of 1025 Hz than expected. To analyze this deviation from the theoretical value and the formula $f_S = |f_2 - f_1|$ experimentally, a short test series is added (see Table 49.1).

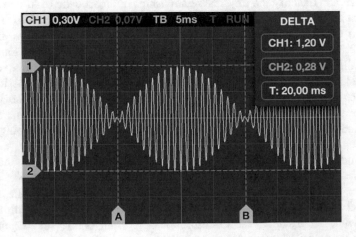

Fig. 49.3 Oscillogram of a beat

Table 49.1 Studying the beat frequency and the perceived tone frequency

f_1 (Hz)	f_2 (Hz)	f_S (Hz)	f_W (Hz)
1000	1050	50.00	1020
		49.75	1042
		50.27	1036
		49.67	1021
	1040	39.83	1024
	1030	29.92	1011
	1020	20.08	1013
	1010	10.12	1003

So the means of the beat frequency f_s average to 49.92 ± 0.27 Hz and the perceived tone frequency to 1028 ± 10.79 Hz, which can be compared to the theoretical values ($f_{S,theo} = 50$ Hz; $f_{W,theo} = 1025$ Hz).

References

1. Kuhn, J., Vogt, P.: Analyzing acoustic phenomena with a smartphone microphone. Phys. Teach. **51**, 118–119 (2013)
2. Schwarz, O., Vogt, P., Kuhn, J.: Acoustic measurements of bouncing balls and the determination of gravitational acceleration. Phys. Teach. **51**, 312–313 (2013)
3. Parolin, S.O., Pezzi, G.: Smartphone-aided measurements of the speed of sound in different gaseous mixtures. Phys. Teach. **51**, 508–509 (2013)
4. Kuhn, J., Vogt, P.: Smartphone & Co. in physics education: effects of learning with new media experimental tools in acoustics. In: Schnotz, W., Kauertz, A., Ludwig, H., Müller, A., Pretsch, J. (eds.) Multidisciplinary Research on Teaching and Learning, pp. 253–269. Palgrave Macmillan, Basingstoke, UK (2015)
5. https://ogy.de/oscilloscope
6. https://ogy.de/audiokit
7. https://ogy.de/frequencygenerator

Cracking Knuckles: A Smartphone Inquiry on Bioacoustics

50

Andreas Müller, Patrik Vogt, Jochen Kuhn, and Marcus Müller

Cracking (or popping) knuckles (or joints) is an (bio-) acoustic phenomenon of which most of you are aware (with dislike, in some cases), and some of you may have wondered where it comes from. We will first give a short explanation, followed by a smartphone experiment validating a central phenomenon of the sound generation mechanism.

50.1 Background

Knuckle cracking sounds originate from special kinds of joints between two bones that have an enclosure (joint capsule) which is filled with a lubricating liquid (synovial fluid). Examples are fingers, toes, elbows, and knees, all of which you may have heard producing the sounds we are interested in here. The enclosure and the liquid within are essential for the sound generation mechanism: When stretching the joint (either by natural movement or by intentional pull), the volume of the joint capsule is increased and the pressure inside thus decreased. But the synovial fluid contains dissolved gas (mainly CO_2), just like water does, and a decrease of pressure on it leads to a partial release of the dissolved gas in the form of bubbles. The bubbles

A. Müller (✉)
Faculty of Science/Physics Section and Institute of Teacher Education, University of Geneva, Geneva, Switzerland
e-mail: Andreas.Mueller@unige.ch

P. Vogt
Institute of Teacher Training (ILF) Mainz, Mainz, Germany
e-mail: vogt@ilf.bildung-rp.de

J. Kuhn · M. Müller
Ludwig-Maximilians-Universität München (LMU Munich), Faculty of Physics, Chair of Physics Education, Munich, Germany
e-mail: jochen.kuhn@lmu.de

© The Author(s), under exclusive license to Springer Nature Switzerland AG 2022
J. Kuhn, P. Vogt (eds.), *Smartphones as Mobile Minilabs in Physics*,
https://doi.org/10.1007/978-3-030-94044-7_50

are visible in radiographs of joints for some time after the cracking occurs (until reabsorption in the liquid). In everyday life, this is well known from opening fizzy drinks; in science, this is described by Henry's law [1]. The sudden generation of a bubble in the elastic medium of the synovial liquid leads to oscillations, in other words, to sound—the one you hear as knuckle cracking [2].

50.2 Experiment

Oscillating bubbles in various liquids are of considerable interest to applied and fundamental science, from early work to understand the pleasant sounds of running water [3] to more recent applications [4] such as medical ultrasound techniques or ocean acoustics, again over a wide range (from habitat research and wildlife preservation to less innocent military purposes). In this strand of research, there is a well-established relationship between oscillation frequency and parameters of the bubble/liquid system [3]:

$$f \approx \frac{1}{2\pi \cdot r} \sqrt{\frac{3p}{\rho}} \ \text{(Minnaert frequency)}, \qquad (50.1)$$

with r being the bubble radius, ρ the density of and p the pressure in surrounding liquid. This result has the familiar square root of "springiness (p)/inertia (ρ)"-structure of the harmonic oscillator frequency formula $\left(f = \frac{1}{2\pi}\sqrt{k/m}\right)$; a simple derivation can be found in Ref. [5]. (In principle, the polytropic index κ appears before p in the above equation, but for an isothermic process it is approximately unity [5]). Solving for r then leads to the basic result of the present experiment:

$$r \approx \frac{1}{2\pi \cdot f} \sqrt{\frac{3p}{\rho}}. \qquad (50.2)$$

This is the place where the portable physics lab offered by smartphones enters, in this case provided by the microphone (we used an iPhone 4), sound storage system, and sound analysis software (app Oscilloscope [6]). With this simple setting, the acoustic signal of a cracking knuckle in Fig. 50.1 was obtained. One clearly sees the expected oscillatory behavior, with an initial transient and then a damped oscillation. From the data in the figure (eight oscillations of 1.5 ms; scale unit is 0.5 ms), one infers a period of $f = 5.3$ kHz.

With approximate input data (f as just inferred; $\rho \approx 1$ kg/L, from water; $p \approx$ some 10^5 Pa (bars), for ambient conditions), one obtains

$$r \approx 0.5\,\text{mm},$$

which is of the order of magnitude of the value $r \approx 0.25$ mm found in the literature [7]. The result of a detailed measurement series is shown in Table 50.1. We

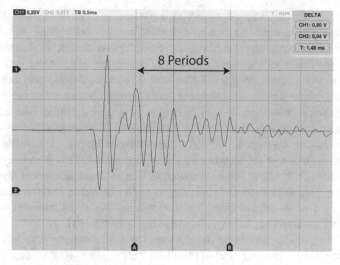

Fig. 50.1 Acoustic signal of a cracking knuckle

Table 50.1 Measurement series on knuckle cracking (index finger; 10 measurements/person); radii according to Eq. (50.2)

Person	Mean frequency f (Hz)	Standard deviation SD (Hz)	Mean radius r (mm)	Standard deviation SD (mm)
1	8813	2003	0.33	0.08
2	5101	1647	0.59	0.16
3	9372	2689	0.32	0.08
4	8251	2493	0.37	0.12
5	8118	2612	0.38	0.11
6	9532	3132	0.33	0.13
7	8551	2166	0.35	0.14
8	7825	1014	0.36	0.04
9	6658	2686	0.47	0.15
10	6882	2838	0.47	0.17

investigated the joint cracking of ten subjects aged between 20 and 30. For each person an index finger was examined ten times. The computed radii varied between 0.3 and 0.6 mm and with an average of 0.396 ($\pm$0.015) mm.

50.3   Comments

In the present discussion about strengths and weaknesses of smartphone science experiments, an often-heard criticism is that many of these experiments are just repetitions of measurements already done with conventional means and that they thus offer nothing really new or interesting. The experiment presented here contributes, by way of example, a relevant argument for this discussion. While it is true that the knuckle cracking experiment could be done with usual sound analysis approaches (e.g., computer based), it is also obvious that it is the mobile and ubiquitous character of the smartphone version of the experiment that makes it accessible to many more people [8]. Many students (or schools) may not have the conventional acoustic analysis equipment, but almost all have a smartphone. Beyond the given example, the experience of having surprising, interesting science (such as visualizing and measuring the cracking gas bubbles in your joint) accessible with one's own means can spark curiosity and self-confidence to investigate other scientific phenomena in the surrounding, everyday world. This then truly becomes "citizen science," [9, 10] which can be seen as one of the objectives of this volume. Furthermore, research shows that studying acoustic phenomena with mobile devices integrated in a more sophisticated instructional setting and combined with more everyday phenomena [11, 12] can also increase learning [13].

Finally, one of course would like to know whether making knuckles crack is harmful; for this issue, see Ref. [14]. The bottom line is that our grandma's exhortations are not justified: you won't get arthritis, and even as many as 30,000 or more "cracking events" during a lifetime don't seem to induce a difference between the hand that was cracked and another that was not.

References

1. Kimbrough, D.R.: Henry's law and noisy knuckles. J. Chem. Educ. **76**, 1509–1510 (1999)
2. Protopapas, M., Cymet, T.: Joint cracking and popping: understanding noises that accompany articular release. J. Am. Osteopath. Assoc. **102**(5), 283–287 (2002)
3. Minnaert, M.: On musical air-bubbles and sounds of running water. Philos. Mag. **16**, 235–248 (1933)
4. Leighton, T.G.: The Acoustic Bubble. Academic, San Diego (1994)
5. Leighton, T.G., Walton, A.J.: An experimental study of the sound emitted from gas bubbles in a liquid. Eur. J. Phys. **8**(2), 98 (1987)
6. https://ogy.de/oscilloscope
7. Courty, J.-M., Kierlik, E.: Craquements de doigts. Pour la Science. **349**, 98–99 (2006)
8. Wong, L.-H., Milrad, M., Specht, M.: Seamless Learning in the Age of Mobile Connectivity. Springer, Dordrecht (2014)
9. Citizen Science Central.: http://www.birds.cornell.edu/citscitoolkit (2015)
10. Fenichel, M., Schweingruber, H.A.: Surrounded by Science: Learning Science in Informal Environments. The National Academies Press, Washington, DC (2010)
11. Vogt, P., Kuhn, J., Neuschwander, D.: Determining ball velocities with smartphones. Phys. Teach. **52**(6), 376–377 (2014)
12. Schwarz, O., Vogt, P., Kuhn, J.: Acoustic measurements of bouncing balls and the determination of gravitational acceleration. Phys. Teach. **51**(5), 312–313 (2013)
13. Kuhn, J., Vogt, P.: Smartphone & Co. in physics education: effects of learning with new media experimental tools in acoustics. In: Schnotz, W., Kauertz, A., Ludwig, H., Müller, A., Pretsch, J. (eds.) Multidisciplinary Research on Teaching and Learning, pp. 253–269. Palgrave Macmillan, Basingstoke, UK (2015)
14. Mirsky, S.: Good news about knuckle cracking. Sci. Am. **25**, 104 (2009)

Shepard Scale Produced and Analyzed with Mobile Devices

51

Kim Ludwig-Petsch and Jochen Kuhn

Contributions in this book describe how smartphones and tablets can be used to analyze experimental data in physics experiments. This is possible in nearly all areas of physics—from classical mechanics (Chaps. 10, 27 and 35) [1–3], to thermodynamics (Chap. 53) [4, 5], electrics (Chap. 56) [6], and radioactivity [7]. While many examples already showed the analysis of acoustic experiments with mobile devices (Chap. 12) [8, 9], no psychoacoustic phenomenon has been studied with the smartphone yet. Nevertheless, acoustic illusions are great starting points for an intrinsic motivation of learners. One great acoustic illusion, often found in science centers, is the so-called Shepard scale illusion. This contribution shows how to use mobile devices to analyze and produce Shepard tones to get a deeper understanding of the acoustic concepts like tone, sound, pitch and frequency.

51.1 Theoretical Background

A Shepard scale is a series of sounds that, when played in a loop, gives the impression of an endless rising scale. It is an auditory illusion, first described by the American psychologist R. Shepard [10] in the 1960s. The individual sounds—the Shepard tones—are periodical signals, which consist of a set of frequencies: the fundamental f_0 plus the harmonics f_n that are powers of two or, in other words, octaves. The amplitude A as a function of time t is:

K. Ludwig-Petsch (✉)
Department of Physics/Physics Education Research Group, Deutsches Museum Munich, Museum Education and Technische Universität Kaiserslautern, Kaiserslautern, Germany
e-mail: k.ludwig-petsch@deutsches-museum.de

J. Kuhn
Ludwig-Maximilians-Universität München (LMU Munich), Faculty of Physics, Chair of Physics Education, Munich, Germany
e-mail: jochen.kuhn@lmu.de

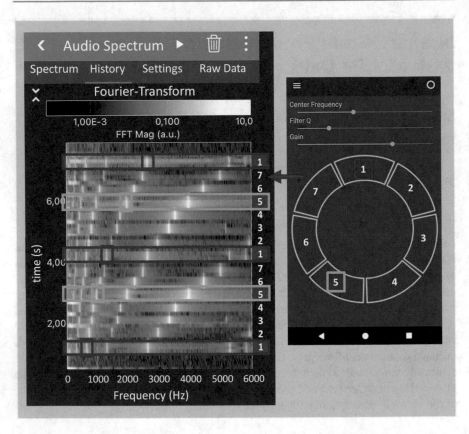

Fig. 51.1 Spectrogram of two full circles of the Shepard scale (tones played one after another). Left: phyphox experiment audio spectrum. Shown is the intensity of the frequencies (white = high, black = low) over the time. The record of the sounds 1 (red) and 5 (green) are highlighted. The spectrogram confirms that the sound of the buttons stays the same. Right: Screenshot of the app Shepard Illusion

$$
\begin{aligned}
A(t) = A_0(\sin 2\pi \cdot f_0 \cdot t) + A_1(\sin 2\pi \cdot f_1 \cdot t) \\
+ A_2(\sin 2\pi \cdot f_2 \cdot t) + \ldots + A_n(\sin 2\pi \cdot f_n \cdot t).
\end{aligned}
\tag{51.1}
$$

For a typical Shepard tone, the amplitudes A_n of the high (e.g. $n > 2$ in Fig. 51.1) and low (e.g. $n < 2$ in Fig. 51.1) frequency harmonics are smaller. This leads to an ambiguity in the perception of the sounds pitch: the same sound can be perceived as low-pitched in the first round and as high-pitched in the second round.

As Suits [11] recently described, the distinction between pitch of a sound and frequency of a sound is important to understand this phenomenon. Many studies show that the perception of pitch is very individual and dependent on the geographical region [12, 13].

In this experiment, students can investigate the number of harmonics of a Shepard tone and compare their frequencies. Furthermore, it can be tested if the harmonic frequencies of a tone are multiples by the power of two of the fundamental:

$$f_0 = \frac{1}{2}f_1 = \frac{1}{4}f_2 = \cdots \frac{1}{2}f_n. \qquad (51.2)$$

51.2 Experimental Setup

Two mobile devices were used to analyze the Shepard scale illusion. The free Android app Shepard Illusion [14] was used on one device to produce the sound—and to explore the illusion. Suits explains how to use audacity on a PC to produce the sounds. On the other device the free application (iOS and Android) phyphox [15] was used to analyze the sounds.

To demonstrate and explore the Shepard illusion with the application in the first step, students can simply push the buttons (1–7, see Fig. 51.1) one after another clockwise to get the illusion of an endless rising scale—or anticlockwise for an endless falling scale. In a second exploration, a stunning comparison between button 1 and 5 can be observed. First step: Playing buttons 1 up to 5 (1, 2, 3, 4, 5), and then alternately 1 and 5, suggests that the pitch of sound 5 is higher than of sound 1. Second step: Playing buttons 5 up to 1 (5, 6, 7, 1), and then alternately 1 and 5, suggests that the pitch of sound 5 is lower than that of sound 1. The illusion of the endless rising scale lies in the perception: The brain is fooled as the perception of pitch of the sound is constantly shifted.

To investigate this phenomenon, the second device is used to record and analyze the Shepard tones. For the analysis the app phyphox was used. It offers the experiment *audio oscilloscope*, which can be used to visualize the sound and show the periodicity of the Shepard tone. The phyphox experiment *audio spectrum* calculates the fast Fourier transform (FFT) and can be used to analyze the spectra of the Shepard tones (Fig. 51.2).

What do the Shepard tones look like on an oscilloscope? Are they really tones or rather a composition of different sinusoidal waves?

51.3 Results

The analysis of the Shepard tones of the app Shepard Illusion confirms that the tones are a periodical signal that is composited of multiple frequencies (Fig. 51.2 shows exemplary sound 1). To analyze the components of the different Shepard tones, we looked at the fast Fourier transform (FFT) of the different tones. The phyphox experiment *audio spectrum* records the sound and calculates the FFT.

According to Eq. (51.2), the lowest detected frequency in the FFT of sound 1 (see Fig. 51.2) is its fundamental f_0 (328 Hz). The other peaks in the FFT are nearly exact

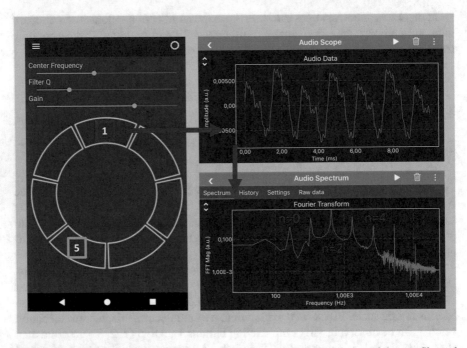

Fig. 51.2 Analysis of the waveform of the Shepard tone 1. *Left*: Screenshot of the app Shepard Illusion. Buttons 1 and 5 are highlighted. *Right*: The oscilloscopic representation reveals that the played notes are not based on a single sinusoidal wave but rather a composition of several frequencies. These can be identified by looking at the audio spectrum (FFT) of the sound. The harmonic frequency with the highest amplitude is $n = 2$

powers of two (656 Hz, 1312 Hz, 2648 Hz, 5273 Hz, 10,546 Hz). This relation can be shown respectively for the sounds 2–7. For example, sound 2 has as its lowest frequency 375 Hz, which is a bit more than one whole step above the fundamental in sound 1. A look at the *History* tab of the experiment *audio spectrum* offers the possibility to record a spectrogram of the Shepard tones. By playing the Shepard tones one after another, a direct comparison is possible. The spectrogram in Fig. 51.1 shows the composition of the single Shepard tones as well as their correlation: One round (1–7) of sounds accords to an octave and is perceived as such.

51.4 Discussion

The Shepard illusion is a great and stunning phenomenon. It combines the physics of acoustics, theory of music, and psychology of perception. The presented experiment offers various options to investigate the science behind this phenomenon with no additional material needed.

Shepard illusions had recently been used in many soundtracks for films, as the illusion of endless rising sounds is a possibility to create tension [16].

References

1. P. Vogt and J. Kuhn, "Analyzing collision processes with the smartphone acceleration sensor," *Phys. Teach.* **52**, 118–119 (Feb. 2014).
2. Becker, S., Klein, P., Kuhn, J.: Video analysis on tablet computers to investigate effects of air resistance. *Phys. Teach.* **54**, 440–441 (2016)
3. J. Kuhn, P. Vogt, and A. Müller, "Analyzing elevator oscillation with the smartphone acceleration sensors," *Phys. Teach.* **52**, 55–56 (Jan. 2014a).
4. M. P. Strzys, S. Kapp, M. Thees, P. Lukowicz, P. Knierim, A. Schmidt, and J. Kuhn, "Augmenting the thermal flux experiment: A mixed reality approach with the HoloLens," *Phys. Teach.* **55**, 376–377 (Sept. 2017).
5. Thees, M., Kapp, S., Strzys, M.P., Beil, F., Lukowicz, P., Kuhn, J.: Effects of augmented reality on learning and cognitive load in university physics laboratory courses. *Comp. Hum. Behav.* **108**, 106316 (2020)
6. S. Kapp et al., "Augmenting Kirchhoff 's laws: Using augmented reality and smart-glasses to enhance conceptual electrical experiments for high school students," *Phys. Teach.* **57**, 52–53 (Jan. 2019).
7. J. Kuhn, A. Molz, S. Gröber, and J. Frübis, "iRadioactivity–Possibilities and limitations for using smartphones and tablet PCs as radioactive counters," *Phys. Teach.* **52**, 351–356 (May 2014b).
8. O. Schwarz, P. Vogt, and J. Kuhn, "Acoustic measurements of bouncing balls and the determination of gravitational acceleration," *Phys. Teach.* **51**, 312–313 (May 2013).
9. Ludwig-Petsch, K., Hirth, M., Kuhn, J.: The sound of a laser blaster: Acoustic dispersion in metal springs analyzed with mobile devices and open source PC audio software. *Phys. Teach.* **60**, 28–33 (Jan. 2022).
10. Shepard, R.N.: Circularity in judgments of relative pitch. *J. Acoust. Soc. Am.* **36**, 2346–2353 (1964)
11. B. H. Suits, "Frequency and pitch," *Phys. Teach.* **57**, 630 (Dec. 2019).
12. Deutsch, D.: Paradoxes of musical pitch. *Sci. Am.* **267**, 88–95 (1992)
13. Deutsch, D.: The processing of pitch combinations. In: *The Psychology of Music*, 3rd edn, pp. 249–325. Elsevier, San Diego (2013)
14. https://ogy.de/shepardillusion
15. https://phyphox.org/
16. Rapan, E.: Shepard tones and production of meaning in recent films: Lucrecia Martel's *Zama* and Christopher Nolan's *Dunkirk*. *The New Soundtrack.* **8**(2), 135–144 (2018)

The Sound of Church Bells: Tracking Down the Secret of a Traditional Arts and Crafts Trade

52

Patrik Vogt, Lutz Kasper, and Jan-Philipp Burde

The sound of church bells is part of most people's everyday life and can easily be examined with smartphones. Similar to other experiments of this book (Chaps. 47 and 49) [1, 2], we use a suitable iOS app. The underlying physical theory of church bells proves to be difficult. A reliable prediction of their natural frequencies based on their exact dimensions is only possible using the finite element method [3]. If you ask bell founders how they calculate the rib of a bell (half longitudinal section of a bell, which completely determines the acoustic properties, Fig. 52.1) in order to get a church bell with the desired frequency spectrum, you will certainly not get an answer: the art of bell casting is based on centuries of experience and knowledge of the rib structure is only shared with direct descendants. We want to have a closer look at these well-guarded secrets, knowing full well that we cannot fully unravel them. This contribution presents simple mathematical models and a comparison with a data set of almost 700 bells.

P. Vogt (✉)
Institute of Teacher Training (ILF) Mainz, Mainz, Germany
e-mail: vogt@ilf.bildung-rp.de

L. Kasper
Department of Physics, University of Education Schwäbisch Gmünd, Schwäbisch Gmünd, Germany
e-mail: lutz.kasper@ph-gmuend.de

J.-P. Burde
University of Tübingen, Physics Education Research Group, Tübingen, Germany
e-mail: Jan-Philipp.Burde@uni-tuebingen.de

Fig. 52.1 Half cross section (rib) of a bell [4]

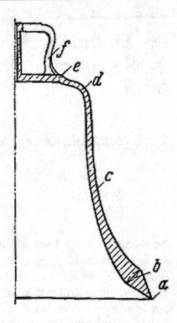

52.1 Vibration Modes of Church Bells

When the clapper strikes against the inside rim of a bell, natural oscillations ensue. As with most musical instruments, this leads to numerous vibrational modes. However, the peculiarity is that the dominant overtones of a bell are not harmonic in the way that stretched strings are. Instead, the frequency spectrum includes pitches that go by the names hum, prime (double the frequency of hum), minor third above that, fifth, and octave (Table 52.1). These intervals, and more, characterize the full, powerful sound of a church bell. In most bells, the perceived strike note roughly corresponds to half the octave frequency and thus approximately coincides with the prime. The root of the frequency spectrum is referred to as the hum, a

Table 52.1 Frequency ratios in the spectrum of a church bell [5]

	Frequency ratio to prime	
Name	Just scale	Church bells
Hum	0.5	0.5
Prime	1.0	1.0
Minor third	1.2	1.183
Quint	1.5	1.506
Octave	2.0	2.0
Major third	2.5	2.514
Fourth	2.667	2.662
Twelfth	3.0	3.011
Upper octave	4.0	4.166

"sub-harmonic" pitch that occurs at half the prime frequency. Its analytical description was an early subject of research in physics, including work by a number of outstanding scientists such as Euler, Jacques Bernoulli, Chladni, Helmholtz, and Rayleigh. Generally, they approached the problem by describing bell-like bodies with significantly simpler geometries, e.g., rings, hemispheres, or hyperboloids [6].

In the next section, we analyze how the hum frequency can be modeled taking into account the bell radius and how the bell radius can be accurately estimated based on a measurement of the hum frequency using a smartphone. In order to verify the new model, a data set with nearly 700 bells, including the hum frequency, bell radius, and the thickness of the inside rim of the bell was created based on a bell book of the Archbishopric of Cologne (Germany) [7]. The idea is to use the extensive experience of bell founders from several centuries to verify and further improve the mathematical model.

52.2   Frequency-Radius Relationship

The data in Fig. 52.2 clearly show a strong relationship between the frequency of the hum tone and the radius or mass, respectively. Initially, rather than attempting to fit this data empirically, we begin the analysis by considering a physically motivated mathematical model.

The equation used by Apfel [8] to model the frequency of wine glasses is a relationship we want to apply to bells, although originally devised for two-dimensional plates bent to a cylinder:

$$f_0 = \frac{\nu_L \cdot d}{\sqrt{3}\pi \cdot R^2} \tag{52.1}$$

(f_0 fundamental frequency, ν_L is the longitudinal or sound velocity in the material, d thickness of the cylinder, R radius). For this purpose, we replace the fundamental

Fig. 52.2 Hum frequency as a function of radius and mass

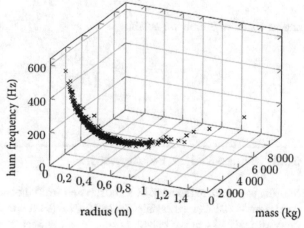

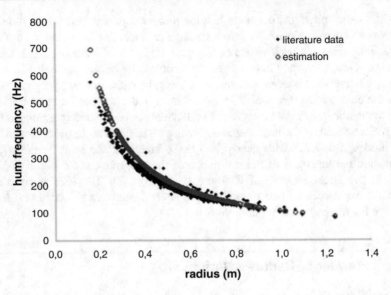

Fig. 52.3 Frequency estimation using a hollow cylinder model vs the actual bell frequencies

frequency f_0 by the hum frequency f_{hum} and we take the speed of sound in bronze for v_L into account ($v_L \approx 3400$ m/s). Furthermore, we also use the relation $d/R \approx 1/7$, which is a result of our own analysis of the data set described in [7]:

$$f_{hum} = \frac{1}{\sqrt{3} \cdot 7 \cdot \pi} \frac{v_L}{R}. \tag{52.2}$$

By introducing a correction factor, the deviation of the data from the model can be reduced to 3.5% on average (Fig. 52.3). We get

$$f_{hum} = 0.092 \cdot \frac{v_L}{\pi \cdot R}, \tag{52.3}$$

and as a rule of thumb we obtain

$$R \approx \frac{100\,\text{Hz}}{f_{hum}}\text{m}. \tag{52.4}$$

52.3   Mass-Radius Relationship

In order to calculate the mass of a church bell from the measured hum frequency, we need a mass-radius relationship. In 1885 Otte [9] found that the mass M of a bell is proportional to its radius cubed, as one might expect from scaling arguments alone.

Based on the data of 700 church bells and the eq. $M = c \cdot R^3$, we empirically find the relationship

$$M = 4776 \frac{\text{kg}}{\text{m}^3} \cdot R^3 \quad \text{and} \tag{52.5}$$

$$M \approx 4776 \frac{\text{kg}}{\text{m}^3} \left(\frac{100 \text{ Hz} \cdot \text{m}}{f_{\text{hum}}} \right)^3 \tag{52.6}$$

With this empirically found mass-radius relationship, the mass of a church bell can easily be estimated based on a frequency measurement with an average deviation of 11.7%.

52.4   Result of a Sample Measurement

With the iOS app Spektroskop [10], a measurement of the "Maria Gloriosa" bell of the Bremen Cathedral (Germany), cast in 1433, was carried out (Fig. 52.4). The bell has a radius of 0.85 m and a mass of 2500 kg [11]. The results show a good approximation of the expected frequency ratios. The hum frequency corresponds to the lowest, large frequency peak at 117 Hz, and the other peaks corresponding to the various overtones can be found in Fig. 52.4. By inserting the measured hum frequency of 117 Hz in relationships (52.4) and (52.6), we get a radius of 0.85 m and

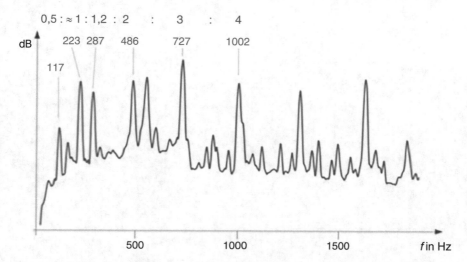

Fig. 52.4 Frequency spectrum of the "Maria Gloriosa" bell of St. Peter's Cathedral, Bremen (Germany), recorded with the app Spektroskop [10] and visualized with a spreadsheet program

a mass of about 2900 kg. While the estimate of the bell radius agrees with the literature value, the mass is overestimated by 16%.

References

1. Kuhn, J., Vogt, P., Hirth, M.: Analyzing the acoustic beat with mobile devices. *Phys. Teach.* **52**, 248–249 (April 2014)
2. Monteiro, M., Marti, A.C., Vogt, P., Kasper, L., Quarthal, D.: Measuring the acoustic response of Helmholtz resonators. *Phys. Teach.* **53**, 247 (April 2015)
3. Perrin, R., Charnley, T.: Normal modes of the modern English church bell. *J. Sound Vib.* **90**(1), 29–49 (1983)
4. Meyer, E., Klaes, J.: Über den Schlagton von Glocken. *Die Naturwissenschaften.* **21**(39), 694–701 (1933)
5. https://en.wikipedia.org/wiki/Strike_tone
6. Lehr, A. "Die Konstruktion von Läuteglocken und Carillonglocken in Vergangenheit und Gegenwart," *Greifenstein* (2005).
7. Hoffs, G., Glocken katholischer Kirchen Kölns (2004). https://ogy.de/Bells
8. Apfel, R.E.: Whispering' waves in a wineglass. *Am. J. Phys.* **53**, 1070–1073 (Nov. 1985)
9. Otte, H.: *Glockenkunde.* T. O. Weigel, Leipzig (1884)
10. https://ogy.de/Spektroskop
11. Data of the Glock available at https://ogy.de/Glockendaten

Part X

Temperature and Heat

Augmenting the Thermal Flux Experiment: A Mixed Reality Approach with the HoloLens

53

M. P. Strzys, S. Kapp, M. Thees, Jochen Kuhn, P. Lukowicz, P. Knierim, and A. Schmidt

In the field of Virtual Reality (VR) and Augmented Reality (AR), technologies have made huge progress during the last years [1–3] and also reached the field of education [4, 5]. The virtuality continuum, ranging from pure virtuality on one side to the real world on the other [6], has been successfully covered by the use of immersive technologies like head-mounted displays, which allow one to embed virtual objects into the real surroundings, leading to a Mixed Reality (MR) experience. In such an environment, digital and real objects do not only coexist, but moreover are also able to interact with each other in real time. These concepts can be used to merge human perception of reality with digitally visualized sensor data, thereby making the invisible visible. As a first example, in this chapter we introduce alongside the basic idea of this column (Chap. 67) [7] an MR experiment in thermodynamics for a laboratory course for freshman students in physics or other science and engineering subjects that uses physical data from mobile devices for analyzing and displaying physical phenomena to students.

M. P. Strzys (✉) · S. Kapp · M. Thees
Department of Physics/Physics Education Group, Technische Universität Kaiserslautern, Kaiserslautern, Germany
e-mail: strzys@physik.uni-kl.de

J. Kuhn
Ludwig-Maximilians-Universität München (LMU Munich), Faculty of Physics, Chair of Physics Education, Munich, Germany
e-mail: jochen.kuhn@lmu.de

P. Lukowicz
German Research Center for Artificial Intelligence (DFKI), Embedded Intelligence Group, Kaiserslautern, Germany

P. Knierim · A. Schmidt
Ludwig-Maximilians-Universität München, Institute of Informatics, Human-Centered Ubiquitous Media, Munich, Germany

53.1 Theoretical Background

The paradigm experiment for heat conduction in metals can be realized with a metallic rod, heated on one side while simultaneously cooled on the other [8]. Our setup allows us to observe the heat flux through the rod, directly on the real physical object, using a false-color representation. Moreover, additional representations such as graphs and numerical values can be included as digital augmentations to the real experiment, allowing for a just-in-time evaluation of physical processes. If the rod is perfectly isolated, after some equilibration time the system will reach a steady state with a hot end at temperature T_1, a cold end at temperature T_2, and a constant spatial gradient along the rod axis, which allows us to calculate the thermal conductivity constant of the material according to

$$\lambda = \frac{L}{A(T_1 - T_2)} \dot{Q},$$

if the constant heating power $\dot{Q}$ and the dimensions, i.e., length L and cross section A, of the rod are known. Here it is assumed that the full heating power $\dot{Q}$ applied to the warm end of the rod will be removed on the cold end by cooling. If the rod is not isolated, it moreover is possible to calculate the loss of heat to the environment according to $h = 2AL$, using the extracted decline factor α from an exponential fit to the experimental data [9].

53.2 Experimental Design

Our heat conduction experiment consists of cylindrical metal sample rods with a length of $L = 26$ cm and a diameter of $d = 5$ cm made of aluminum and copper, respectively (Fig. 53.1b). The sample is heated at one end with a cartridge heater and cooled at the other by a standard CPU fan. The isolated version moreover has a PVC insulation layer with a 3-mm slit along the rod, to allow for thermal imaging of the rod inside (Fig. 53.1c). The temperature data finally are extracted from thermal images taken with an infrared camera placed in front of the sample. Each pixel of the image along a fixed line in axial direction yields one temperature value, such that the whole spatial distribution can be captured by a single shot. The data are then passed to the Microsoft HoloLens via WiFi, where the visualization is done. In the current state, an app provides three different representations of the data: false-color image of the temperature values projected as a "hologram" [10] directly onto the sample cylinder, numerical values at three predefined points, and a temperature graph as a function of the position along the rod hovering above the setup, cf. Figure 53.1a. The user may switch the numerical and graph representation on and off at will; moreover, it is possible to export the data as a CSV file at any time for later analysis. These functions can be executed with the help of virtual buttons (cf. the three white squares in Fig. 53.1a) projected at the right end of the rod, which can be selected by the gaze and triggered by hand gestures.

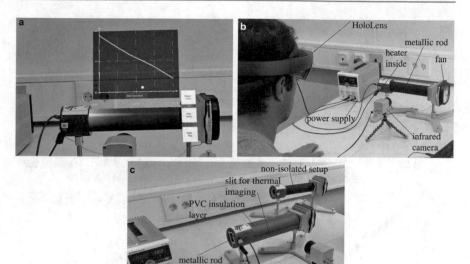

Fig. 53.1 (**a**) Experimental setup (non-isolated) with MR experience; augmented representations: false-color representation of temperature along the rod, numerical values at three points above the rod, temperature graph. (**b**) Experimental setup (non-isolated) and user wearing a HoloLens. (**c**) Isolated setup

The benefit of the MR experiment setup is the possibility of keeping track of the real physical devices and representations of the data simultaneously and in real time, as the representations are continuously updated with new data from the camera. The false-color representations allow one to experience an otherwise invisible quantity, like in this case temperature, with human senses, thus extending perception to new regimes. Furthermore, direct feedback is implemented, such that students get an immediate impression of effects of the experimental parameters.

53.3 Experimental Results

The exported data can be analyzed using standard instruments like spreadsheet programs. In the isolated case, the problem reduces to a linear fit of the temperature graph, yielding the slope and thus, with the help of the constant heating power $\dot{Q} = 50$ W used in the experiments via (1), the thermal conductivity constant λ. In our tests (Figs. 53.2(a and b)) the isolated setup yielded a value of $\lambda_{Al} = (128 \pm 11)$ W/(m·K) for the aluminum rod and $\lambda_{Cu} = (329 \pm 27)$ W/(m·K) for the copper rod. As expected, due to the nonperfect isolation of the samples, these values are smaller than the reference values found in the literature, $\lambda_{Al,lit} = 235$ W/(m·K) and $\lambda_{Cu,lit} = 401$ W/(m·K), as the effective heat flux $\dot{Q}$ along the rod is reduced by loss to the environment.

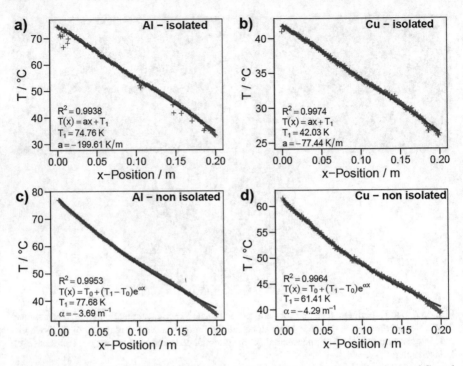

Fig. 53.2 Temperature graph after equilibration and linear fit for an isolated Al rod (**a**) and Cu rod (**b**). Temperature graph after equilibration and exponential fit for a non-isolated Al rod (**c**) and Cu rod (**d**)

With the help of the non-isolated rod, the heat transfer coefficient h can be determined using the exponential decline factor α (Figs. 53.2(c and d)), yielding $h_{Al} = (0.72 \pm 0.01)$ W/K and $h_{Cu} = (2.5 \pm 0.01)$ W/K, respectively. These values are also underestimated for the same reasons; however, compared to $h_{Al, lit} = (1.32 \pm 0.01)$ W/K and $h_{Cu,lit} = (3.05 \pm 0.01)$ W/K, where the literature values for λ were used, they are reasonable regarding the very simple insulation of the setup.

53.4 Conclusion

The MR experimental setup for a thermal flux experiment presented here sheds new light on an experiment well known in physics laboratory courses. The main focus of this design is to visualize the invisible and thus to extend human perception to new regimes, e.g., temperature and heat, thereby strengthening the connection between theory and experiment. In this realization the MR setting not only has the advantage of intrinsic contextuality, but also spatial and time contiguity, which is supposed to support the learning process of the students [11, 12]. Moreover, the just-in-time evaluation of the data yields the possibility for the students to directly examine the process itself and the parameter involved, and immediately compare the outcome to theoretical predictions, which we believe to enhance the links between theory and experiment. Under that perspective the effort to achieve possibly more exact numerical values for quantities like the thermal conductivity therefore seem to be less important in this setting. Instead, the technical support during the experimental phase will give students the possibility to thoroughly examine the relationship between cause and effect and thus deepen their physical understanding.

Research shows that learning with MR glasses in this topic could reduce cognitive load and increase conceptual understanding [13, 14]. Current developments extend this idea to other topics which have to be studied further as they show more contradictory results [15–17].

Support from the German Federal Ministry of Education and Research (BMBF) via the project "Be-greifen" is gratefully acknowledged [18].

References

1. Schmalstieg, D., Höllerer, T.: *Augmented Reality: Principles and Practice*. Addison-Wesley Professional (2016)
2. Sandor, C., Fuchs, M., Cassinelli, A., Li, H., Newcombe, R.A., Yamamoto, G., Feiner, S.K.: Breaking the barriers to true augmented reality. *CoRR*. abs/1512.05471 (2015)
3. Hockett, P., Ingleby, T.: Augmented reality with HoloLens: experiential architectures embedded in the real world. *CoRR*. abs/1610.04281 (2016)
4. Santos, M.E.C., Chen, A., Taketomi, T., Yamamoto, G., Miyazaki, J., Kato, H.: Augmented reality learning experiences: survey of prototype design and evaluation. *IEEE Trans. Learn. Technol.* 7(1), 38–56 (Jan. 2014)
5. Kuhn, J., Lukowicz, P., Hirth, M., Poxrucker, A., Weppner, J., Younas, J.: gPhysics–Using smart glasses for head-centered, context-aware learning in physics experiments. *IEEE Trans. Learn. Technol.* 9(4), 304–317 (2016)
6. Milgram, P., Kishino, F.: A taxonomy of mixed reality visual displays. *IEICE Trans. Info. Systems.* 77(12), 1321–1329 (1994)
7. Kuhn, J., Vogt, P.: Diffraction experiments with infrared remote controls. *Phys. Teach.* 50, 118 (Feb. 2012)
8. Parrot, J.E., Stuckes, A.D.: *Thermal Conductivity of Solids*. Pion Limited, London (1975)
9. See additional online material for a derivation and more detailed discussion at *TPT Online* at 10.1119/1.4999739 under the SUPPLEMENTAL tab.
10. The HoloLens does not actually use interference-based holograms but rather projections to the transparent head mounted displays of the device. However, the term hologram is used by the manufacturer to describe these.
11. Crawford, M.L.: Teaching Contextually: Research, Rationale, and Techniques for Improving Student Motivation and Achievement in Mathematics and Science. CORD (CCI Publishing) (2001)
12. Mayer, R.E. (ed.): *The Cambridge Handbook of Multimedia Learning*. Cambridge University Press (2010)
13. Strzys, M.P., Kapp, S., Thees, M., Klein, P., Lukowicz, P., Knierim, P., Schmidt, A., Kuhn, J.: Physics holo.lab learning experience: Using Smartglasses for Augmented Reality labwork to foster the concepts of heat conduction. *Eur. J Phys.* 39(3), 035703 (2018)
14. Thees, M., Kapp, S., Strzys, M.P., Beil, F., Lukowicz, P., Kuhn, J.: Effects of augmented reality on learning and cognitive load in university physics laboratory courses. *Comp. Hum. Behav.* 108, 106316 (2020)

15. Kapp, S., Thees, M., Strzys, M.P., Beil, F., Kuhn, J., Amiraslanov, O., Javaheri, H.M., Lukowicz, P., Lauer, F., Rheinländer, C., Wehn, N.: Augmenting Kirchhoff's laws: using augmented reality and smartglasses to enhance conceptual electrical experiments for high school students. *Phys. Teach.* **57**(1), 52–53 (2019)
16. Donhauser, A., Küchemann, S., Rau, M., Malone, S., Edelsbrunner, P., Lichtenberger, A., Kuhn, J.: Making the invisible visible: visualization of the connection between magnetic field, electric current and Lorentz force with the help of Augmented Reality. *Phys. Teach.* **58**(6), 438–439 (2020)
17. Altmeyer, K., Kapp, S., Thees, M., Malone, S., Kuhn, J., Brünken, R.: Augmented reality to foster conceptual knowledge acquisition in STEM laboratory courses–theoretical derivations and empirical findings. *Brit. J Educ. Technol.* **51**(3), 611–628 (2020)
18. The Microsoft HoloLens could be purchased for $3000 USD. https://www.microsoft.com/en-us/hololens/buy

Smartphones: Experiments with an External Thermistor Circuit

54

Kyle Forinash and Raymond F. Wisman

This contribution adds a further example to illustrate how to use the headphone port of a smartphone to receive data from an external circuit, in this case, a simple, adaptable homemade example for temperature measurement (Chap. 58) [1].[1]

Temperature is routinely measured by a thermistor in a simple circuit. Figures 54.1 and 54.2 show such a circuit that can be controlled by an iPod, iPad, iPhone or Android app. Many physics students are familiar with the usual setup for this circuit where a dc current is fed to the circuit and the voltage across the thermistor is monitored [2]. A change in temperature changes the resistance of the thermistor, which can be recorded as a dc voltage change (such as by an Apple II).

Thermistor resistance, R, is exponentially related to temperature by the equation $R = Ae^{\frac{B}{T}}$, where A and B are constants. Taking a natural log of both sides gives the linear equation $\ln(R) = \ln(A) + \frac{B}{T}$. The constants are determined by calibrating with a few known temperatures.

To repeat this measurement using the headphone port on an iPhone or iPod requires just a little more work. Designed for audio, the headphone port filters dc current inputs and outputs. However, a sine wave with frequency ranging up to 20 kHz can be output by a phone app from the headphone port to serve as input to the temperature circuit. The amplitude of the returning sine wave output is determined by the resistance of the thermistor. Only the amplitude of the sine wave varies in the circuit, not the frequency. The device returns V/V_0 in decibels (dB), where V is the voltage measured between ring three (the microphone input) and the tip of the headphone jack (Fig. 54.2) and V_0 is an internal reference voltage. The output dB reading is proportional to $\ln(R)$ since $R = V/i$. The amplitude of the circuit output is

---

[1] An audio jack adaptor can be used if the iPhone or iPad does not have a headphone jack.

K. Forinash (✉) · R. F. Wisman
Emeritus, School of Natural Sciences, Indiana University Southeast, New Albany, IN, USA
e-mail: kforinas@ius.edu; rwisman@ius.edu

Fig. 54.1 Measuring
temperature using an external
circuit and iPod

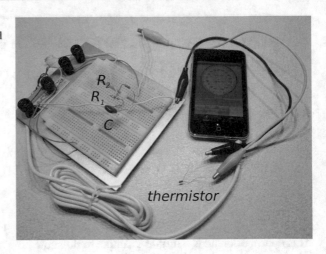

Fig. 54.2 Circuit diagram for
Fig. 54.1. $R_1 = 10$ kΩ,
$R_2 = 220$ Ω, and $C = 0.1$ μF.
The headphone jack is a
standard four-pole jack

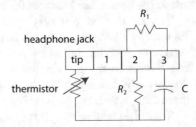

measured through the headphone microphone jack as a decibel (dB) reading by our
phone app [3]. Plotting dB for several independently measured temperatures and
fitting the plot to a straight line gives the constants A and B (which contain the
constant reference voltage V_0 and constant current I). With the constants A and B,
determined from the plot (Fig. 54.3), the circuit is calibrated to measure other
temperatures. To reduce noise due to sampling errors, a software low pass filter is
applied to the peak decibel reading of the returning sine wave as measured over a
short period of time. Programmed with the equation, the app can measure temperature changes, then email or plot results (Fig. 54.4). The headphone volume control
serves to calibrate for individual circuit differences.

Surprisingly, the thermistor is very sensitive to small temperature changes (air
currents, for example). Encasing the thermistor in something with a little thermal
inertia (for example, the shell of a plastic ink pen) adds some stability to the
measurements.

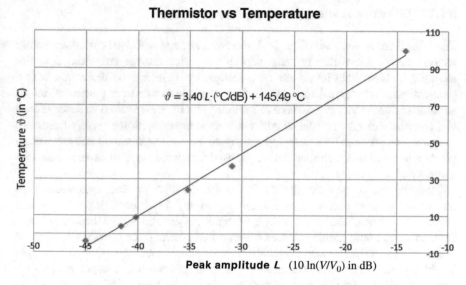

Thermistor vs Temperature

$\vartheta = 3.40\,L\cdot(°C/dB) + 145.49\,°C$

Temperature ϑ (in °C)

Peak amplitude L $(10\,\ln(V/V_0)$ in dB$)$

Fig. 54.3 As explained in the text, the iPhone returns V/V_0 in decibels, dB, which is proportional to $\ln(R)$. The plot is of temperature (in °C) versus peak amplitude L from the thermistor in dB taken for temperatures, respectively, in the freezer compartment of a refrigerator, in ice water, in the refrigerator, in an air-conditioned room, outside during the day, and in boiling water

Fig. 54.4 The thermistor app plotting temperature changes over time

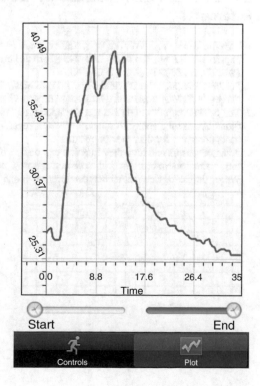

54.1   Other measurements

The same thermistor circuit (Figs. 54.1 and 54.2) and app can also be used with other measurement devices that produce similar resistance change characteristics. For example, ambient light levels can be measured by replacing the thermistor with a light-activated photo resistor. As another example, some water sensors act as variable resistors that are sensitive to humidity. Strain gauges used in force probes and elsewhere change resistance when subjected to stress. Some types of pressure gauges signal pressure by a changing resistance. Basically, the circuit and app turns a phone into a measurement tool that can be used for a wide range of experiments with a variety of sensors [4].

Although the above examples are for an iPhone, iPad, or iPod, the concepts are not limited to these particular mobile devices as any smartphone or programmable device with a headphone could, with appropriate programming, use the same circuit. Data collection with mobile devices offers a vast potential for science experiments and is at the same beginning stage that computer data collection was when the Apple II with a game port was introduced some 40 years ago. As early computers opened a whole range of scientific experiments to a wider audience and inspired a new generation of scientists, we feel smartphones and other mobile devices offer comparable possibilities.

References

1. For other examples, see Forinash, K. and Wisman, R. "Smartphones as portable oscilloscopes for physics labs," *Phys. Teach.* **50**, 242–243 (April 2012) and "Photogate timing with a smartphone," *Phys. Teach.* **53**, 234 (April 2015).
2. See, for example, John W. Snider and Joseph Priest, *Electronics for Physics Experiments: Using the Apple II Computer/Book with Disk* (Addison-Wesley, Reading, 1989).
3. The original app, "Mobile Science–Temperature," is no longer available from the Apple App Store. Contact the author, Raymond Wisman, regarding implementation questions on using the headphone port for data measurement.
4. A list of sensors and how they function can be found at https://www.sensorland.com. Ready to run external thermometers with software are now commercially available, for example from Weber (https://ogy.de/ambienttemperatureprobe) and Thermodo (http://thermodo.com).

Studying Cooling Curves with a Smartphone

55

Manuela Ramos Silva, Pablo Martín-Ramos, and Pedro Pereira da Silva

This contribution describes a simple procedure for the study of the cooling of a spherical body using a standard thermometer and a smartphone. Experiments making use of smartphone sensors have been described before, contributing to an improved teaching of classical mechanics (Chaps. 8, 13, 27 and 29) [1–11], but rarely expand to thermodynamics (Chaps. 53 and 54) [12–14]. In this experiment, instead of using a smartphone camera to slow down a fast movement, we are using the device to speed up a slow process. For that we propose the use of the free app Framelapse [15] to take periodic pictures (in the form of a time-lapse video) and then the free app VidAnalysis [16] to track the position of the mercury inside the thermometer, thus effortlessly tracking the temperature of a cooling body (Fig. 55.1).

The experiment consists of filling a round-bottom flask (five flasks with standard sizes—50, 100, 250, 500, and 1000 mL—were borrowed from the chemistry lab) with hot water, placing a mercury thermometer in the opening and taking periodic pictures of it. The Framelapse app allows the user to set the time interval between pictures (30 s was found to be a suitable choice) and it is a way of automatically monitoring a lengthy experiment (ca. two hours for the 1-L flask).

The video is then processed in the smartphone using VidAnalysis. This intuitive and easy-to-use app requires the setting of the axes, a length scale—which can be done by using the thermometer scale—and the tracking of the mercury position through screen touching (Fig. 55.2, right), frame by frame, generating temperature-vs.-time graphs. The data can then be exported into a CSV file. The file can be further

M. R. Silva (✉) · P. P. da Silva
CFisUC, Department of Physics, FCTUC, Universidade de Coimbra, Coimbra, Portugal
e-mail: manuela@uc.pt; psidonio@uc.pt

P. Martín-Ramos
CFisUC, Department of Physics, FCTUC, Universidade de Coimbra, Coimbra, Portugal

EPS, Universidad de Zaragoza, Huesca, Spain
e-mail: pmr@unizar.es

Fig. 55.1 Photograph of the
experimental setup

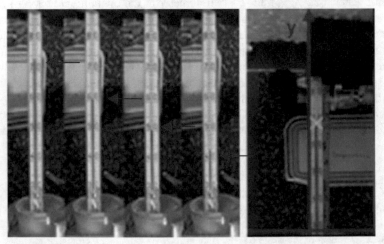

Fig. 55.2 Some of the frames taken during the cooling of the hot water with Framelapse app (left)
and mercury position tracking with VidAnalysis app (right)

manipulated to allow logarithmization or fitting of the experimental curves directly in the smartphone using the Google Sheets app (other popular free apps cannot open CSV files) or in a computer with OpenOffice Calc or MS Excel.

55.1 Newton's Law of Cooling

The temperature of a hot object placed in a cooler surrounding will slowly decrease until it matches that of the environment. It decreases by a combination of three phenomena: conduction, convection, and radiation. The heat flow coming from conduction and convection depends linearly on the difference of the temperature of the object and that of its surroundings [17]:

$$\frac{dQ}{dt} \propto \left(T_{object} - T_{env}\right). \tag{55.1}$$

On the other hand, the heat flow coming from radiation is ruled by the Stefan-Boltzmann law, with a dependence on the fourth power of T. The net heat flow from the object radiation and the surrounding radiation is linearly dependent on ΔT only for very small temperature differences.

Anyhow, for cases where the radiative processes are not predominant, as the one discussed herein, Newton's cooling law can be used to model the system:

$$T_{object}(t) = T_{env} + \left(T_{object}(0) - T_{env}\right) \cdot \exp\left(-t/\tau\right) \tag{55.2}$$

with

$$\tau = Mc/hS, \tag{55.3}$$

M being the mass of the body, S the surface area, c the specific heat capacity per unit mass, and h the convective heat transfer parameter.

Figure 55.3 shows the variation of the temperature of our flasks filled with water; a least-squares fit using a simple exponential (LOGEST function in Google Sheets app) yields $T_{env} = 25.21(5)$ °C, $T_{object}(0) - T_{env} = 55.72$ (4) °C and 3877(8) s for the decay time constant.

55.2 Linear Dependence on M/S Quotient

By applying logarithms on both sides of Eq. (55.2), a linear dependence is obtained:

$$\ln\left(T_{object} - T_{env}\right) = \left(T_{object}(0) - T_{env}\right) - t/\tau. \tag{55.4}$$

A fit of the plotted data (using the LINEST function) yields the slope value, that is, the reciprocal of τ. On its own, τ depends on four parameters (Eq. (55.3)): c is kept constant and h is also approximately constant (since the conditions of the experiment

Fig. 55.3 Plot of decay of water temperature as a function of time for the five round-bottom flask

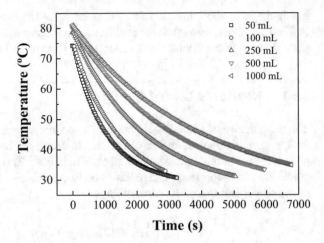

Table 55.1 Mass, outer surface, and decay constant values for the five runs

Flask	M (kg)	R (cm)	S (m^2)	M/S (kg·m^{-2})	τ (s)
#1	0.056	2.45	0.0075	7.42	1294.46
#2	0.115	3.13	0.0123	9.37	1509.87
#3	0.275	4.25	0.0227	12.12	2159.60
#4	0.530	5.25	0.0346	15.30	2983.80
#5	1.052	6.75	0.0573	18.37	3926.34

were kept as close to each other as possible: a fan at medium speed was kept in a corner of the room to keep the ambient air conditions as similar as possible in all the runs) [18]. The other two parameters, M and S, were changing in the five performed runs. Their quotient depends only on the radius of the outer surface:

$$\frac{M}{S} = \frac{\rho \frac{4}{3}\pi R^3}{4\pi R^2} = \frac{\rho}{3}R, \qquad (55.5)$$

where ρ is the density of water and R is the radius. The τ values obtained using this procedure are summarized in Table 55.1 and plotted in Fig. 55.4.

The slope of the τ vs. M/S linear fit (with $R^2 = 0.99$) yields the constant $c/h = 244.6$ m^{2}s/kg, or using $c = 4.186$ J/(g · °C), one gets $h \approx 58$ W/(m^2·°C), which is in the expected range according to Ref. [17]. h is dependent on the geometry, flow orientation, surface roughness, and combination of material/fluid.

The proposed experiment can be useful for the introduction of exponential functions at introductory levels, and can easily be changed to investigate the heat capacity of materials by letting blocks of different materials cool in air or water [19–21]. Since the procedure can readily be adapted to allow the simultaneous tracking of several thermometers, it may also prove useful for investigating the cooling of a nonhomogeneous body (e.g., simulating the estimation of the time since death of a human body [22], if our students are CSI fans!). This experiment benefits from the

Fig. 55.4 Plot of as a function of M/S quotient (or R, for simplicity we have just used the radius of the flask) with a least-squares linear fit

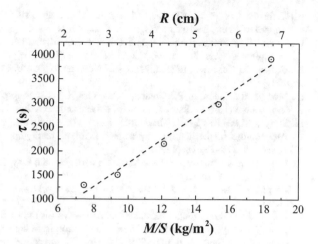

small size and portability of the setup, from running autonomously once set (freeing both students and teachers to other tasks) and from using a device that students bring voluntarily to class (although a sophisticated camera plus a computer can replace the smartphone). Profiting from the above characteristics, more complex experiments in non-ambient conditions (e.g., inside a fridge) can also be envisaged.

Acknowledgments CFisUC gratefully acknowledges funding from FCT Portugal through Grant UID/FIS/04564/2016. P.M-R would like to thank Santander Universidades for its financial support through the Becas Iberoamérica Jóvenes Profesores e Investigadores, España 2017 scholarship program.

References

1. Chevrier, J., Madani, L., Ledenmat, S., Bsiesy, A.: Teaching classical mechanics using smartphones. *Phys. Teach.* **51**, 376 (Sept. 2013)
2. Hall, J.: More smartphone acceleration. *Phys. Teach.* **51**, 6 (Jan. 2013)
3. Monteiro, M., Stari, C., Cabeza, C., Marti, A.C.: The Atwood machine revisited using smartphones. *Phys. Teach.* **53**, 373–374 (Sept. 2015)
4. Shakur, A., Kraft, J.: Measurement of Coriolis acceleration with a smartphone. *Phys. Teach.* **54**, 288–290 (May 2016)
5. Vogt, P., Kuhn, J.: Analyzing simple pendulum phenomena with a smartphone acceleration sensor. *Phys. Teach.* **50**, 439 (Oct. 2012)
6. Vogt, P., Kuhn, J., Müller, S.: Experiments using cell phones in physics classroom education: The computer-aided g determination. *Phys. Teach.* **49**, 383–384 (Sept. 2011)
7. Becker, S., Klein, P., Kuhn, J.: Video analysis on tablet computers to investigate effects of air resistance. *Phys. Teach.* **54**, 440–441 (Oct. 2016)
8. Gröber, S., Klein, P., Kuhn, J.: Video-based problems in introductory mechanics physics courses. *Eur. J. Phys.* **35**(5), 055019 (2014)
9. Klein, P., Kuhn, J., Müller, A., Gröber, S.: Video Analysis Exercises in Regular Introductory Physics Courses: Effects of Conventional Methods and Possibilities of Mobile Devices. In: Schnotz, W., Kauertz, A., Ludwig, H., Müller, A., Pretsch, J. (eds.) *Multidisciplinary Research on Teaching and Learning*, pp. 270–288. Palgrave Macmillan UK, London (2015)

10. Klein, P., Gröber, S., Kuhn, J., Müller, A.: Video analysis of projectile motion using tablet computers as experimental tools. *Phys. Educ.* **49**(1), 37–40 (2014)
11. Pereira, V., Martín-Ramos, P., da Silva, P.P., Silva, M.R.: Studying 3D collisions with smartphones. *Phys. Teach.* **55**, 312–313 (May 2017)
12. Brown, D., Cox, A.J.: Innovative uses of video analysis. *Phys. Teach.* **47**(3), 145–150 (March 2009)
13. Moggio, L., Onorato, P., Gratton, L.M., Oss, S.: Time-lapse and slow-motion tracking of temperature changes: Response time of a thermometer. *Phys. Educ.* **52**(2), 023005 (2017)
14. Strzys, M.P., Kapp, S., Thees, M., Kuhn, J., Lukowicz, P., Knierim, P., Schmidt, A.: Augmenting the thermal flux experiment: A mixed reality approach with the HoloLens. *Phys. Teach.* **55**, 376–377 (Sept. 2017)
15. Neximo Labs, Framelapse–Time Lapse Camera (Google Play, 2017). https://ogy.de/Timelapse
16. https://ogy.de/VidAnalysis
17. Bergman, T.L., Lavine, A.: *Fundamentals of Heat and Mass Transfer*, 8th edn. Wiley, Inc, Hoboken, NJ (2017)
18. This is a simplified model for the actual pathways for the transfer of energy. In reality, three major pathways come into play: the free convection from the warm water to the glass flask, the conduction through the flask walls, and the forced convection from the flask to the room air. For advanced classes, all 3 phenomena can be taken in consideration and their relative effect pondered by changing the fan speed.
19. Mattos, C.R., Gaspar, A.: Introducing specific heat through cooling curves. *Phys. Teach.* **40**, 415–416 (Oct. 2002)
20. Planinšič, G., Vollmer, M.: The surface-to-volume ratio in thermal physics: From cheese cube physics to animal metabolism. *Eur. J. Phys.* **29**(2), 369–384 (2008)
21. Will, J.B., Kruyt, N.P., Venner, C.H.: An experimental study of forced convective heat transfer from smooth, solid spheres. *Int. J. Heat Mass Transfer.* **109**, 1059–1067 (2017)
22. Leinbach, C.: Beyond Newton's law of cooling – Estimation of time since death. *Int. J. Math. Educ. Sci. Technol.* **42**(6), 765–774 (2011)

Part XI

Electricity and Magnetism

Real-time Visualization of Electrical Circuit Schematics: An Augmented Reality Experiment Setup to Foster Representational Knowledge in Introductory Physics Education

56

Luisa Lauer, Markus Peschel, Sarah Malone, Kristin Altmeyer, Roland Brünken, Hamraz Javaheri, Orkhan Amiraslanov, Agnes Grünerbl, and Paul Lukowicz

Empirical research has shown that augmented reality (AR) has the potential to promote learning in different contexts [1, 2]. In particular, this has been shown for AR-supported physics experiments, where virtual elements (e.g., measurement data) were integrated into the learners' visual reality in real-time: compared to traditional experimentation, AR reduced cognitive load [3] and promoted conceptual learning [4, 5]. Drawing upon previous work [6], we present an AR-supported experiment on simple electrical circuits that allows for real-time visualization including highlighting of electrical circuit schematics using either smartglasses or tablet computers. The experiment addresses students in introductory physics education and holds the potential to provide visual assistance for complex electrical circuits in secondary or higher physics education.

L. Lauer (✉) · M. Peschel
Department of Physics, Saarland University, Saarbrücken, Germany
e-mail: luisa.lauer@uni-saarland.de; markus.peschel@uni-saarland.de

S. Malone · K. Altmeyer · R. Brünken
Department of Education, Saarland University, Saarbrücken, Germany
e-mail: s.malone@mx.uni-saarland.de; kristin.altmeyer@uni-saarland.de;
r.bruenken@mx.uni-saarland.de

H. Javaheri · O. Amiraslanov · A. Grünerbl · P. Lukowicz
German Research Center for Artificial Intelligence (DFKI), Embedded Intelligence Group,
Kaiserslautern, Germany
e-mail: hamraz.javaheri@dfki.de; agnes.gruenerbl@dfki.de; paul.lukowicz@dfki.de

© The Author(s), under exclusive license to Springer Nature Switzerland AG 2022
J. Kuhn, P. Vogt (eds.), *Smartphones as Mobile Minilabs in Physics*,
https://doi.org/10.1007/978-3-030-94044-7_56

56.1   Theoretical Background

Representational competencies—the ability to use different domain-specific representations and to effortlessly switch between them—is considered an important skill in science instruction that is closely related to the acquisition of conceptual knowledge [7]. However, in introductory physics education, the use of symbolic representations for electrical circuits can evoke difficulties and thus lead to learning impairment, especially when the spatial arrangement of the circuit components does not resemble the clear structure of the corresponding circuit schematic [8]. The subsequently presented experiment may facilitate the introduction to electrical circuits for students in lower physics or even primary science education and improve their future conceptual learning.

56.2   Experiment Setup

We use the Microsoft HoloLens [9] as smartglasses (Fig. 56.1) and a Samsung Galaxy [9] tablet computer (Fig. 56.2) to supplement the students' field of view with real-time visualizations of electrical circuit schematics corresponding to actual circuits (or parts of circuits) in order to augment the students' perception of real objects with their correct symbolic representation. As the smartglasses are securely mounted on the students' heads (Fig. 56.1) and the tablet computer is fastened with a

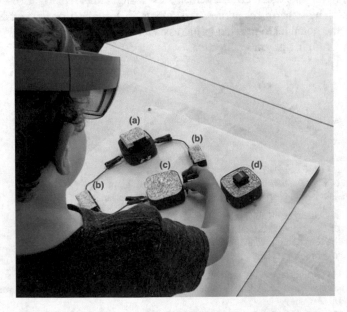

Fig. 56.1 A student building a circuit with the HoloLens setup using battery (**a**), cables (**b**), light bulb (**c**) and switch (**d**)

Fig. 56.2 A student exploring the tools while building an electrical circuit with the tablet setup. Once a tool is touched, the corresponding circuit symbol is highlighted

bracket (Fig. 56.2), the students can build and modify the circuits with their hands free and experience the immediate adaption of the AR-circuit schematic.

An application specifically developed for this purpose displays the circuit schematic symbols above single circuit components as well as full schematics of electrical circuits and allows the circuit schematic to successively assemble in real-time with the construction of the electrical circuit (Figs. 56.2 and 56.3). To strengthen the cognitive association between the component and its symbolic representation, each time a certain component is touched or slightly moved, the corresponding AR-schematic symbol is highlighted (Fig. 56.2). To implement the described real-time adaption of the AR-visualization, we integrated a cable identification system into the circuit components as suggested in previous work [5] and a touch/movement sensor system that both transmit data to the AR-application via a wireless network connection.

A combination of RFID tracking and marker-based tracking installed in the circuit components enables the displayed circuit schematics or symbols to follow the position of the components; while the RFID tracking localizes the horizontal position of the components in relation to a reference array mat, the marker-based tracking improves the vertical localization, e.g., when students lift the components from the table.

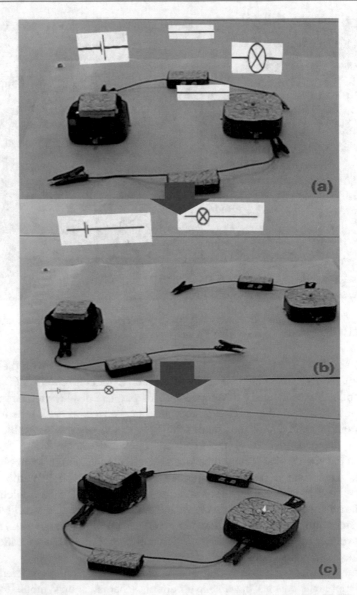

Fig. 56.3 Snapshots from the HoloLens during a successive circuit assembly: (**a**) Virtual circuit symbols are displayed above the single circuit components. (**b**) As components are connected, a virtual circuit schematic assembles. (**c**) As a full circuit is set up, the virtual circuit schematic adapts accordingly

56.3 Experiment Procedure

The experiment consists of two phases. In the familiarizing phase, the students are introduced to the circuit components, their purpose in an electrical circuit, and their circuit symbols displayed by AR. Likewise, they are accustomed to the touch-highlighting of the AR-circuit symbols and the merging of symbols into a circuit schematic when components are connected.

During the subsequent work phase, the students are guided to build several simple electrical circuits using cables, light bulbs, and a switch while seeing the corresponding real-time AR-circuit schematics.

56.4 Outlook on Further Developments

The presented AR experiment allows for encountering learning difficulties concerning simple electrical circuits in introductory physics courses from primary to secondary education. By adapting the hard- and software for detection and visualization of parallel circuits, the experiment could be used in the curriculum of electrical circuits in physics education from introductory level up to higher courses. Moreover, the AR-highlighting function holds potential for easing the understanding of highly complex electrical circuits, which may be beneficial for conceptual knowledge acquisition, especially for secondary or higher physics education.

Acknowledgments The work described is funded by the German Federal Ministry of Education and Research.

References

1. Garzón, J., Acevedo, J.: Meta-analysis of the impact of augmented reality on students' learning gains. *Educ. Res. Rev.* **27**, 244–260 (2019)
2. Akçayır, M., Akçayır, G.: Advantages and challenges associated with augmented reality for education: A systematic review of the literature. *Educ. Res. Rev.* **20**, 1–11 (2017)
3. Thees, M., Kapp, S., Strzys, M., Beil, F., Lukowicz, P., Kuhn, J.: Effects of augmented reality on learning and cognitive load in university physics laboratory courses. *Comp. Human Behav.* **108**, 106316 (2020)
4. Altmeyer, K., Kapp, S., Thees, M., Malone, S., Kuhn, J., Brünken, R.: The use of augmented reality to foster conceptual knowledge acquisition in STEM laboratory courses–Theoretical background and empirical results. *Brit. J Educ. Tech.* **51**(3), 611–628 (2020)
5. Strzys, M.P., Kapp, S., Thees, M., Klein, P., Lukowicz, P., Knierim, P., Schmidt, A., Kuhn, J.: Physics holo.lab learning experience: Using Smartglasses for Augmented Reality labwork to foster the concepts of heat conduction. *Eur. J Phys.* **39**(3), 035703 (2018)

6. Kapp, S., Thees, M., Strzys, M.P., Beil, F., Kuhn, J., Amiraslanov, O., Javaheri, H., Lukowicz, P., Lauer, F., Rheinländer, C., Wehn, N.: Augmenting Kirchhoff 's laws: Using augmented reality and smartglasses to enhance conceptual electrical experiments for high school students. *Phys. Teach.* **57**, 52–53 (2019)
7. Hubber, P., Tytler, R., Haslam, F.: Teaching and learning about force with a representational focus: Pedagogy and teacher change. *Res. Sci. Educ.* **40**(1), 5–28 (2010)
8. Wilhelm, T., Hopf, M.: Schülervorstellungen zum elektrischen Stromkreis. In: Schecker, H., Wilhelm, T., Hopf, M., Duit, R. (eds.) *Schülervorstellungen und Physikunterricht*, pp. 115–138. Springer Spektrum, Berlin (2018)
9. This publication is neither affiliated with, nor sponsored or approved by, Microsoft Corp. or Samsung Corp.

Augmenting Kirchhoff's Laws: Using Augmented Reality and Smartglasses to Enhance Conceptual Electrical Experiments for High School Students

57

Sebastian Kapp, Michael Thees, Martin P. Strzys, Fabian Beil,
Jochen Kuhn, Orkhan Amiraslanov, Hamraz Javaheri,
Paul Lukowicz, Frederik Lauer, Carl Rheinländer, and Norbert Wehn

During the last decade the development of modern digital media such as smartphones and tablet computers has enabled new experimental possibilities in STEM education. Besides these now nearly ubiquitous devices, the fields of virtual reality (VR) and augmented reality (AR) also made huge progress [1] and reached education [2, 3]. In this chapter we introduce an AR experiment alongside the basic idea of iPhysicsLabs-column [4] and following prior work [5–7]. In the experiment high school students use smartglasses and real-time measurement data to study Kirchhoff's circuit laws in electrical DC circuits.

57.1 Theoretical Background

(a) *Kirchhoff's first law:* The total current as the sum of linear superposed single currents meeting at a node is zero.

S. Kapp (✉) · M. Thees · M. P. Strzys · F. Beil
Department of Physics/Physics Education Group, Technische Universität Kaiserslautern,
Kaiserslautern, Germany
e-mail: kapp@physik.uni-kl.de

J. Kuhn
Ludwig-Maximilians-Universität München (LMU Munich), Faculty of Physics, Chair of Physics
Education, Munich, Germany
e-mail: jochen.kuhn@lmu.de

O. Amiraslanov · H. Javaheri · P. Lukowicz
German Research Center for Artificial Intelligence (DFKI), Embedded Intelligence Group,
Kaiserslautern, Germany

F. Lauer · C. Rheinländer · N. Wehn
Department of Electrical and Computer Engineering, Technische Universität Kaiserslautern,
Microelectronic Systems Design Research Group, Kaiserslautern, Germany

J. Kuhn, P. Vogt (eds.), *Smartphones as Mobile Minilabs in Physics*,
https://doi.org/10.1007/978-3-030-94044-7_57

This superposition respects electrical current as a signed quantity according to the flow towards or away from the node. It is based on the principle of conservation of electric charge and the repulsive Coulomb forces originating from the charges. It can be stated as

$$I_{\text{total}} = \sum_{k=1}^{n} I_k = 0,$$

with n representing the total number of branches.

(b) ***Kirchhoff's second law:*** The total voltage as the sum of all (signed) electrical potential differences around any closed network is zero.

This law requires the directed measurement of voltages and is based on the principle of conservation of energy. Similar to (a), it can be stated as

$$U_{\text{total}} = \sum_{k=1}^{n} U_k = 0,$$

with n as the total number of (signed) voltages measured.

57.2   Experiment Setup and Procedure

We use the Microsoft HoloLens [8] as AR smartglasses to merge human perception of reality with digitally visualized real-time sensor data into the user's field of view. Thus, we obtain spatial and temporal contiguity between reality and its representations to reduce a spatial and temporal split of the learner's attention while performing the experiment [9]. To do so, we enhance common experimental boxes with sensor modules that contain digital current and voltage meters, a processing unit, and a wireless interface to transfer the sensor data in real time to the smartglasses. There, the data are processed into different representational forms (e.g., line graphs, bar plots, numerical values, etc.) that are presented on the device's transparent displays. Using visual markers on the sensor boxes, these visualizations can then be spatially pinned to the corresponding box. Still having the hands free, the learner can continue to manipulate the experimental setup and immediately see the consequent measurement data without breaking the experimental flow by iteratively measuring isolated data points.

Equipped with this AR system and the enhanced experimental equipment, the students study Kirchhoff's circuit laws in two steps:

(a) First, they connect three different electrical resistors according to given schematic diagrams to get first impressions on how to conduct these experiments and to be focused on the differences between sequential and parallel setups (Fig. 57.1(a and b)).

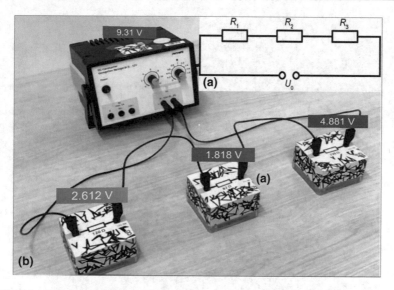

Fig. 57.1 Introductory part of the experiment: (**a**) schematic diagram, (**b**) view through the smartglasses

(b) Given four different resistors, students then are able to evaluate the concepts they built up in part (a) by comparing their expectations with the measurement data from more complex, self-designed circuits (Figs. 57.2(a and b)). This might lead to conflicts with their predictions and a reconsideration of the already built-up mental models. Thus, students are engaged to explore Kirchhoff's laws by creating their own experiments and probing their hypotheses.

57.3   Results and Outlook

Once a circuit is closed by the students, the sensor data are visualized in their field of view via the smartglasses; meanwhile, they can choose whether electrical current or voltage is displayed. Currently, the data are presented as numerical values or as an amplitude of a virtual pointer instrument to allow quantitative as well as qualitative analysis of the circuits. Furthermore, the students can focus on the conceptual evaluation of electrical circuits without the need to perform a repetitive measurement while still working with real setups and data.

In the next step, each box will be equipped with an RFID reader in its bottom, which allows us to locate the sensor boxes on an array of RFID tags under the surface of the table. With this, the system can spatially pin the correct visualizations to the corresponding box without the need for distracting visual markers. Furthermore, we will add a cable identification system to the experimental boxes, revealing their interconnections. To do so, we use modified connectors for the cables (like those

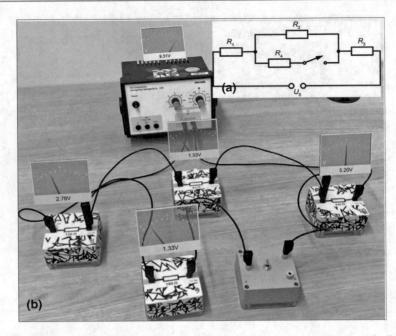

Fig. 57.2 Advanced part of the experiment, view through the smartglasses: (**a**) schematic diagram, (**b**) visualization as a pointer instrument

from stereo headphone jacks), devoting one connection to the electrical circuit and the other two to the identification system. Each connector is equipped with a specific unique resistor between the identification pins to be separated from other cables. Using this information, we are able to highlight errors in the experimental setup and guide the students through complex circuits. In addition, we can visualize common model representations as additional simulations according to the real-time data (see Fig. 57.3).

Finally, the usage of the whole tool kit as an assisting system for explorative learning settings could be a big step towards addressing common misconceptions in electricity. While research shows that learning heat conduction with AR smartglasses in this topic could reduce cognitive load and increase conceptual understanding [6–11]. Current developments extend this idea to other topics which have to be studied further as they show more contradictory results [12, 13].

Fig. 57.3 Simple visualization of real-time data (voltage) according to the electron gas model [10]

References

1. Schmalstieg, D., Höllerer, T.: *Augmented Reality: Principles and Practice.* Addison-Wesley Professional (2016)
2. Santos, M.E.C., Chen, A., Taketomi, T., Yamamoto, G., Mi-yazaki, J., Kato, H.: Augmented reality learning experiences: Survey of prototype design and evaluation. *IEEE Trans. Learn. Technol.* **7**(1), 38 (Jan. 2014)
3. Kuhn, J., Lukowicz, P., Hirth, M., Poxrucker, A., Weppner, J., Younas, J.: gPhysics – Using smart glasses for head-centered, context-aware learning in physics experiments. *IEEE Trans. Learn. Technol.* **9**(4), 304 (2016)
4. Kuhn, J., Vogt, P.: Diffraction experiments with infrared remote controls. Phys. Teach. **50**, 118 (Feb. 2012)
5. Strzys, M.P., Kapp, S., Thees, M., Klein, P., Lukowicz, P., Knierim, P., Schmidt, A., Kuhn, J.: Physics holo.lab learning experience: Using smart glasses for augmented reality lab work to foster the concepts of heat conduction. *Eur. J. Phys.* **39**(3), 035703 (2018)
6. Strzys, M.P., Thees, M., Kapp, S., Klein, P., Knierim, P., Schmidt, A., Lukowicz, P., Kuhn, J.: Smartglasses as Assistive Tools for Undergraduate and Introductory STEM Laboratory Courses. In: Buchem, I., Klamma, R., Wild, F. (eds.) *Perspectives on Wearable Enhanced Learning: Current Trends, Research and Practice.* Springer, Heidelberg (2018)
7. Strzys, M.P., Kapp, S., Thees, M., Kuhn, J., Lukowicz, P., Knierim, P., Schmidt, A.: Augmenting the thermal flux experiment: A mixed reality approach with the Holo Lens. *Phys. Teach.* **55**, 376 (Sept. 2017)
8. The Microsoft HoloLens Development Edition can be ordered for $3000 USD at https://www.microsoft.com/en-us/hololens/buy. (This is an independent publication and is neither affiliated with, nor authorized, sponsored, or approved by, Microsoft Corporation).
9. Moreno, R.: Learning in high-tech and multimedia environments. *Curr. Psychol. Sci.* **15**(2), 63 (2006)

10. Burde, J.P., Wilhelm, T.: Concept and empirical evaluation of a new curriculum to teach electricity with a focus on voltage. *PERC Proceedings*. **2017**, 68 (2017)
11. Thees, M., Kapp, S., Strzys, M.P., Beil, F., Lukowicz, P., Kuhn, J.: Effects of augmented reality on learning and cognitive load in university physics laboratory courses. *Comp. Hum. Behav.* **108**, 106316 (2020)
12. Donhauser, A., Küchemann, S., Rau, M., Malone, S., Edelsbrunner, P., Lichtenberger, A., Kuhn, J.: Making the invisible visible: Visualization of the connection between magnetic field, electric current and Lorentz force with the help of Augmented Reality. *Phys. Teach.* **58**(6), 438–439 (2020)
13. Altmeyer, K., Kapp, S., Thees, M., Malone, S., Kuhn, J., Brünken, R.: Augmented Reality to Foster Conceptual Knowledge Acquisition in STEM Laboratory Courses–Theoretical Derivations and Empirical Findings. *Brit. J Educ. Technol.* **51**(3), 611–628 (2020)

Smartphones as Portable Oscilloscopes for Physics Labs

58

Kyle Forinash and Raymond F. Wisman

Given that today's smartphones are mobile and have more computing power and means to measure the external world than early PCs, they may also revolutionize data collection, both in structured physics laboratory settings and in less predictable situations, outside the classroom. Several examples using the internal sensors available in a smartphone were presented in earlier chapters in this book (Chaps. 6 and 67) [1, 2]. But data collection is not limited only to the phone's internal sensors since most also have a headphone port for connecting an external microphone and speakers. This port can be used to connect to external equipment in much the same way as the game port on the early Apple II was used in school labs. Below is an illustration using the headphone port to receive data from an external circuit: smartphones as a portable oscilloscope using commercially available hardware and applications.

Plugging an oscilloscope probe, offered by the German company HMB-TEC [3], into the headphone port and downloading software available from several sources [4], turns an iPhone, iPad [5] or Android device into an oscilloscope. Fig. 58.1 shows the probe and SignalScope software in action. Any sound analysis app that allows input from the microphone jack and shows the raw signal will work.

The use of the headphone port does impose limitations on oscilloscope measurements. Although the audio hardware in most mobile devices is designed for sample rates as high as 44.1 kHz, the audio port frequency response range for earlier iPhones is between 100 Hz and 8 kHz. The iPhone 3G and later models have a fairly flat frequency response between 10 Hz and about 20 kHz, and can accept up to a 5-V signal (although lower voltages are recommended). A further limitation of using an iPhone as an oscilloscope is that, unlike an oscilloscope that allows dc and low frequencies to be measured, the iPhone uses ac coupling to connect to the microphone. This means that low frequencies are filtered, which will distort a square

K. Forinash (✉) · R. F. Wisman
Emeritus, School of Natural Sciences, Indiana University Southeast, New Albany, IN, USA
e-mail: kforinas@ius.edu; rwisman@ius.edu

© The Author(s), under exclusive license to Springer Nature Switzerland AG 2022
J. Kuhn, P. Vogt (eds.), *Smartphones as Mobile Minilabs in Physics*,
https://doi.org/10.1007/978-3-030-94044-7_58

Fig. 58.1 iPhone as oscilloscope using the SignalScope app. The oscilloscope probe is plugged into the iPhone headphone port and connected across a resistor in a series *LRC* circuit driven by a sine wave. The circuit elements are a 1000-Ω resistor, a 25-mH inductor, and a 0.1-μF capacitor

Fig. 58.2 Circuit diagram for Fig. 58.1. The oscilloscope probes are placed across the resistor

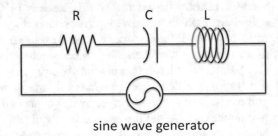

sine wave generator

wave signal somewhat. However we found that sine waves up to at least 12,000 Hz are accurately represented without distortion on an iPhone 3GS and iPhone 7.

As a trial experiment we used the iPhone and oscilloscope app to determine the resonance frequency of a standard series combination of an inductor, resistor, and capacitor (*LRC*). The circuit (Figs. 58.1 and 58.2) has a 25-mH inductor, a 1000-W resistor, and a 0.1-mF capacitor in series driven by a sine wave generator. Figure 58.3 shows a graph of the amplitude of the signal, measured by the iPhone, across the resistor as a function of frequency.

The resonance frequency

$$f = \frac{1}{2\pi}\sqrt{\frac{1}{LC}}$$

is about 3200 Hz. SignalScope is actually a bit easier to use than an oscilloscope because there are cross hairs that can be used to pick out a data point, the value of which appears on the screen.

SignalScope can be calibrated to an external source (for example a set voltage) and set to appropriate units (since the port was designed to work with a microphone, the default vertical scale unit is the pascal). The time scale can be set between 0.1 ms and 500 ms by a two-finger pinch-to-expand command; the amplitude scale is

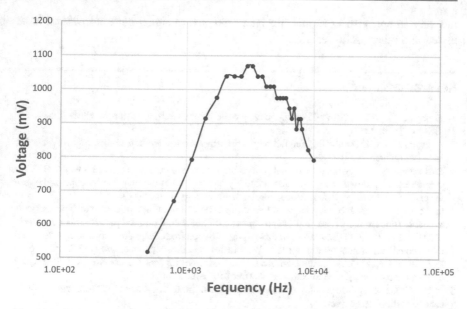

Fig. 58.3 Voltage versus frequency taken from SignalScope on the iPhone across the resistor of an *LRC* circuit driven by a sine wave at different frequencies. Resonance occurs at 3200 Hz

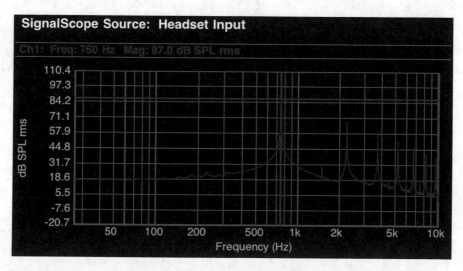

Fig. 58.4 SignalScope app screenshot of the fast Fourier transform (FFT) of a triangle wave of 750 Hz from a wave generator

adjusted similarly. Screenshots can be saved as pictures and a trigger mode captures transient voltages. A real-time fast Fourier transform option has several choices of windowing, averaging, and frequency resolution (Fig. 58.4). The vertical axis can be logarithmic or decibel. The SignalScope Pro 2020 version adds the options of saving raw data in spreadsheet format (for later analysis), a signal generator, sound level meter, and other useful features.

References

1. Kuhn, J., Vogt, P.: Diffraction experiments with infrared remote controls. *Phys. Teach.* **50**, 118–119 (Feb. 2012)
2. Vogt, P., Kuhn, J.: Analyzing free fall with a smartphone acceleration sensor. *Phys. Teach.* **50**, 144–145 (March 2012)
3. This company no longer sells external probes but a standard oscilloscope probe with a BNC to phone jack adaptor will work. It is also relatively easy to build your own oscilloscope probe (see for example https://ogy.de/oscilloscopeprobe or https://ogy.de/tastkopf).
4. In the original version of this chapter we used the free version of SignalScope which no longer is available (nor are the other oscilloscope apps mentioned in the original paper). There are Basic ($39.99) and Pro 2020 ($299.99) versions of SignalScope from Faber Acoustical (https://ogy.de/faberacoustical) which we did not test. We did test Oscilloscope (https://ogy.de/oscilloscope) from ONYX Apps ($9.99), Audio Kit from Sinusoid (free) for Mac OS and SmartScope (https://ogy.de/smartscope), from LabNation tools (free) for Android OS.
5. If the iPhone or iPad does not have a headphone jack, the oscilloscope probe will work with an iPhone audio jack adaptor.

The Flashing Light Bulb: A Quantitative Introduction to the Theory of Alternating Current

59

Patrik Vogt, Stefan Küchemann, and Jochen Kuhn

In this book several previous chapters focused on mechanics experiments that can be analyzed using a mobile video motion analysis (Chaps. 3, 11, 13, 27 and 36) [1–5]. However, the use of this method is also possible in completely different areas, which is the focus of this contribution.

Based on a high-speed recording of an incandescent light bulb, which shows periodic fluctuations in the brightness that cannot be perceived by the eye under normal conditions, the question is raised regarding the origin of these fluctuations. Here we show that the AC voltage of the power grid is the cause for these fluctuations, and careful observation allows a determination of the power line frequency [6]. Since the brightness of an incandescent lamp in a simple resistive circuit is independent of the direction of the current, the bulb is equally bright whether the voltage is positive or negative. Therefore, to calculate the power line frequency, the determined flashing frequency must be halved (Fig. 59.1).

59.1 Why Alternating Voltage and Determination of the Power Line Frequency?

The main reason that alternating voltage is used in power grids for energy transmission is the transformability of the voltage. In order to avoid heat loss, the voltage of long-distance lines is transformed to high values. In common households, however, a significantly lower and less dangerous voltage is used. If there is a need for a higher

P. Vogt (✉)
Institute of Teacher Training (ILF) Mainz, Mainz, Germany
e-mail: vogt@ilf.bildung-rp.de

S. Küchemann · J. Kuhn
Ludwig-Maximilians-Universität München (LMU Munich), Faculty of Physics, Chair of Physics Education, Munich, Germany
e-mail: jochen.kuhn@lmu.de

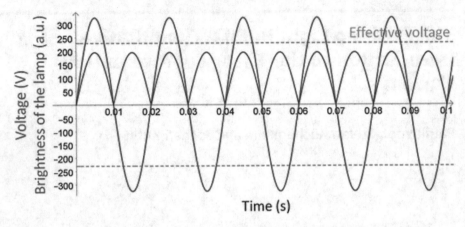

Fig. 59.1 Relationship between voltage and brightness (for a frequency of 50 Hz)

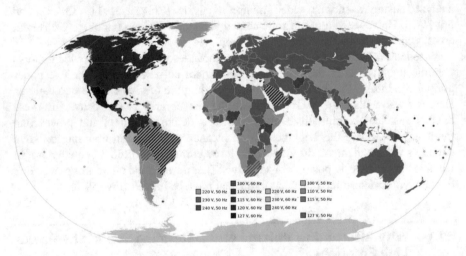

Fig. 59.2 Worldwide distribution of the power line frequency (Wikipedia, https://ogy.de/ACVoltage)

voltage in a household, the voltage can be transformed using electrical transformers. The power line frequency in most parts of the world is 50 Hz and in Northern America it is 60 Hz (Fig. 59.2). The frequency value is not arbitrary, but follows several criteria:[1] Higher frequencies allow smaller transformer cores (lighter and cheaper), but they also produce larger conduction losses due to the skin effect and cause larger phase shifts as the frequency is directly related to the number of rotations and poles of generators and motors (increase of centrifugal forces). Therefore, the choice of 50 or 60 Hz is a compromise of various factors.

---

[1] "Netzfrequenz", Wikipedia, https://de.wikipedia.org/wiki/Netzfrequenz

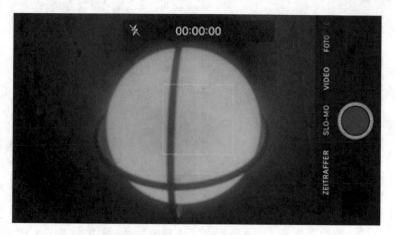

Fig. 59.3 Filming an incandescent lamp in slow-motion mode

59.2 Procedure

The investigated experiment has already been described in Ref. [7], but the authors used a digital camera with a frame rate of 1000 fps. In this work, we used a conventional smartphone, which allows students to perform the experiment at home without the need for additional equipment. New smartphone models typically feature a high-speed recording mode but the maximum frame rate is often limited to 240 fps. The significantly lower number of frames per second in comparison to the experiment in Ref. [7] requires an additional reduction in the playback speed for video analysis. This can easily be done directly in the smartphone using free apps.

In this experiment, we record an incandescent lamp (alternatively, LED lamps can be used if they are not operated via a stabilized DC voltage) in slow-motion mode (Fig. 59.3 for a snapshot). The frame rate was set to the maximum value of 240 fps in the device settings of the iPhone 6 s used here.

59.3 Analysis and Results

When recording at 240 fps, the playback speed is 1/8 of the actual process speed. Here, the periodic changes in brightness are already clearly visible but they are still too fast for a quantitative evaluation. Therefore, we further reduce the playback speed with an appropriate app. In the example measurement described here, the application VivaVideo[2] was used for this purpose. The app allows a reduction of the playback speed by a factor of 0.25. The output video was saved and its playback speed was additionally reduced by a factor of 0.25. Consequently, the actual process

---

[2]https://ogy.de/iOSVivaVideo (iOS), https://ogy.de/VivaVideoAndroid (Android).

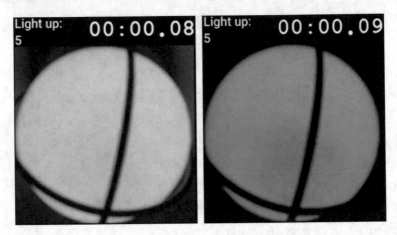

Fig. 59.4 Two snapshots taken from the video (left at maximum brightness, right at a lower brightness level)

is 128 times faster compared to the playback speed of the slow-motion video. Finally, the slow-motion video was cut to a length of exactly one minute, which implies that the displayed process takes 60/128 ≈ 0.47 s in real time (Fig. 59.4).[3] In this video, the lamp lights up 47 times, which results in an actual blinking frequency of 100 Hz (= 47/60 · 128 Hz) and a power line frequency of 50 Hz, corresponding exactly to the frequency of the European power grid. Since the filament always "afterglows" a little, the lamp does not turn off completely. However, a complete turn off can be achieved using a light-emitting diode that is connected to a non-stabilized DC voltage source and operated slightly above the threshold voltage.[4]

59.4 Summary

The experiment described in this chapter shows that video motion analysis could not only be used to described movements of objects but also in completely different areas. The analysis of periodic fluctuations of a light bulb can be used as an introduction to the theory of alternating current. For students, the flashing of the light bulb might be surprising at first and it can only be explained by a periodically changing voltage. The use of smartphones as experimental tool can be helpful during learning. As research shows, learning with mobile video analysis could increase conceptual understanding [8, 9] while decreasing irrelevant cognitive effort and negative emotions [8, 10].

---

[3] The video "flashing light bulb" starts at maximum brightness and then shows the bulb lighting up 47 more times; https://ogy.de/chap59-video1

[4] The video "LED" shows a blinking LED recorded via Casio EX-ZR400 (1000 fps); https://ogy.de/chap59-video2

References

1. Becker, S., Klein, P., Kuhn, J.: Video analysis on tablet computers to investigate effects of air resistance. *Phys. Teach.* **54**, 440–441 (2016)
2. Pereira, V., Martin-Ramos, P., Pereira Da Silva, P., Ramos Silva, M.: Studying 3D collisions with smartphones. *Phys. Teach.* **55**, 312–313 (2017)
3. Thees, M., Becker, S., Rexigel, E., Cullmann, N., Kuhn, J.: Coupled pendulums on a clothesline. *Phys. Teach.* **56**, 404–405 (2018)
4. Becker, S., Thees, M., Kuhn, J.: The dynamics of the magnetic linear accelerator examined by video motion analysis. *Phys. Teach.* **56**, 484–485 (2018)
5. Priyanto, A., Aji, Y., Aji, M.P.: An experiment of relative velocity in a train using a smartphone. *Phys. Teach.* **58**, 72–73 (Feb. 2020)
6. Planinšič, G.: Fluorescence and phosphorescence: Easier to investigate than to spell. *Phys. Teach.* **54**, 442 (Oct. 2016)
7. Kapser, L., Vogt, P.: "30 × 45 Minuten Physik", in *Fertige Stundenbilder für Highlights zwischendurch* (Klasse 7-10). Verlag an der Ruhr, Mühlheim (2015)
8. Becker, S., Gößling, A., Klein, P., Kuhn, J.: Using mobile devices to enhance inquiry-based learning processes. *Learn. Instr.* **69**(2020), 101350 (2020a)
9. Hochberg, K., Becker, S., Louis, M., Klein, P., Kuhn, J.: Using smartphones as experimental tools–A follow-up: Cognitive effects by video analysis and reduction of cognitive load by multiple representations. *J. Sci. Educ. Technol.* **29**(2), 303–317 (2020)
10. Becker, S., Gößling, A., Klein, P., Kuhn, J.: Investigating dynamic visualizations of multiple representations using mobile video analysis in physics lessons: Effects on emotion, cognitive load and conceptual understanding. *Zeitschrift für Didaktik der Naturwissenschaften.* **26**(1), 123–142 (2020b)

Observation of the Magnetic Field using a Smartphone

60

Yasuo Ogawara, Shovit Bhari, and Steve Mahrley

When measuring the magnetic field using a sensor in a smartphone, one can choose from several applications [1–3] with useful graphical interfaces. For my first endeavor, I tried to locate the magnetic field sensor (MFS) in my smartphone SC-04D. I was able to find plenty of pictures on the internet of my phone dismantled, but none of them helped me locate the MFS. I did not want to risk dismantling my own phone, so I used a piece of weakly magnetized iron and found a point where the magnetic field reading was the highest (188 μT) with the application Sensor Box for Android [4]. I then assumed the sensor was just below the surface of that point (Fig. 60.1). In an earlier paper [5], Silva introduced readers to the app 3D Compass and Magnetometer [6] and "Relative Mode," which can be used to "zero out" background noise such as geomagnetism. Some [7] similar applications available record data that can be exported as a CSV file and can be analyzed using MS Excel or other spreadsheet programs.

60.1 Magnetic Field of a Circular Electric Current

The magnetic field of a long straight wire can be measured using a compass, but it is not so easy to measure the magnetic field produced by a circular electric current. However, utilizing my smartphone made it somewhat easy with the application 3D Compass and Magnetometer [8] (Fig. 60.2).

Y. Ogawara (✉)
Keio Senior High School, Kanagawa, Yokohama, Japan
e-mail: ogawara@hs.keio.ac.jp

S. Bhari · S. Mahrley
Department of Physics, California State University Fullerton, Fullerton, CA, USA

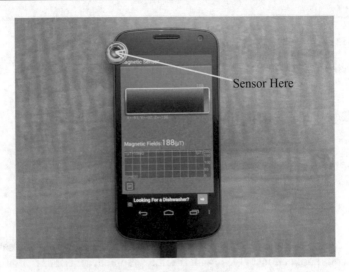

Fig. 60.1 Finding the location of the magnetic field sensor

Fig. 60.2 Measuring the magnetic field of a circular electric current

The following,

$$B = \mu_0 \frac{I}{2r}$$

and

Table 60.1 Electric Current and Measurements of the magnetic field

| | $z = 0$ m | | $z = 0.07$ m | |
| | Theory | Measure ments | Theory | Measure ments |
I (A)	B (μT)	B (μT)	B (μT)	B (μT)
0	0	0	0	0
0.5	3.9	3.9	1.7	1.9
1.0	7.9	7.9	3.3	3.2
1.5	11.8	11.5	5.0	4.5
2.0	15.7	15.1	6.7	6.3
2.5	19.6	19.0	8.4	7.9

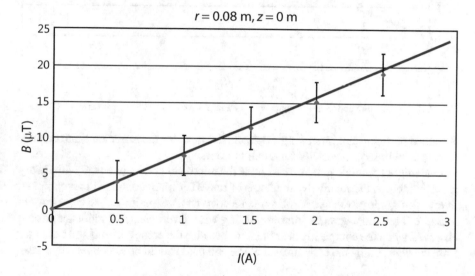

Fig. 60.3 Theory line and measurements with standard error ($r = 0.08$ m, $z = 0$ m)

$$B = \mu_0 \frac{Ir^2}{2(r^2 + z^2)^{3/2}},$$

is the accepted theory [9] to calculate the z-component of the magnetic field of a circular electric current, at its center and on its axis, respectively.

60.2 Results

When I used the radius of the circular current to be $r = 0.08$ m and the distance from its center to the sensor to be z as in Fig. 60.2, I obtained the measurements shown here (Table 60.1 and Figs. 60.3 and 60.4). Even though the circle I made was not exact and the current-carrying wires next to the circle (and a power supply) likely

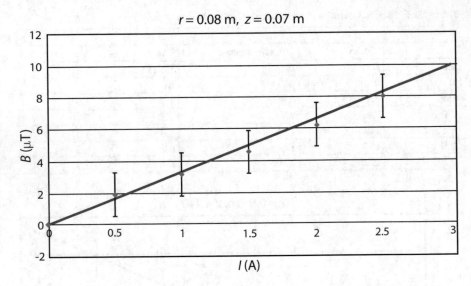

Fig. 60.4 Theory line and measurements with standard error ($r = 0.08$ m, $z = 0.07$ m)

produced non-constant interfering magnetic fields, these measurements agreed very closely with theory, especially for small currents.

One might speculate that the reason for the relatively good agreement with theory is attributable to the relatively small size of the MFS when compared to a compass. In my 12th-grade physics class, all students conduct an experiment in a laboratory to measure **B** around a straight current-carrying wire. Every year, a significant error is observed when the compass is near the wire because the compass tip is still rather far from the wire. Using the MFS instead of the compass should be effective in this case.

Acknowledgments Yasuo Ogawara wrote this chapter while in residence at the Catalyst Center at California State University Fullerton as a visiting scholar. Yasuo appreciates Dr. Michael Loverude with respect to have accepted and mentioned some important points in physics education.

Reference

1. https://ogy.de/sensorboxandroid
2. https://ogy.de/3dcompassmagnetometer
3. https://ogy.de/sensortesterapk
4. https://ogy.de/sensorboxandroid
5. Silva, N.: Magnetic field sensor. *Phys. Teach.* **50**(372) (2012)
6. https://ogy.de/3dcompassmagnetometer
7. https://ogy.de/sensortesterapk
8. https://ogy.de/3dcompassmagnetometer
9. http://hyperphysics.phy-astr.gsu.edu/hbase/magnetic/curloo.html

Magnetic Fields Produced by Electric Railways

61

Martín Monteiro, Giovanni Organtini, and Arturo C. Martí

We propose a simple experiment to explore magnetic fields created by electric railways and compare them with a simple model and parameters estimated using easily available information. A pedestrian walking on an overpass above train tracks registers the components of the magnetic field with the built-in magnetometer of a smartphone. The experimental results are successfully compared with a model of the magnetic field of the transmission lines and the local Earth's magnetic field. This experiment, suitable for a field trip, involves several abilities, such as modeling the magnetic field of power lines, looking up reliable information, and estimating non-easily accessible quantities.

61.1 Electromagnetic Fields

We live surrounded by electromagnetic fields covering all ranges of spatial and temporal scales. Most of them are difficult to measure or even to detect. Notable exceptions are static or quasi-static magnetic fields, which can be detected or measured with a compass or with a smartphone sensor. In the past, smartphone experiments were proposed using small currents (Chap. 62) [1–3], magnets [4, 5], and the Earth magnetic field [6]. The other example, not so obvious, consists in the magnetic fields produced by power electric currents. Most of the power systems are

M. Monteiro (✉)
Universidad ORT Uruguay, Montevideo, Uruguay
e-mail: monteiro@ort.edu.uy

G. Organtini
Universidad di Roma La Sapienza, Rome, Italy
e-mail: Giovanni.Organtini@roma1.infn.it

A. C. Martí
Universidad de la República, Montevideo, Uruguay
e-mail: marti@fisica.edu.uy

© The Author(s), under exclusive license to Springer Nature Switzerland AG 2022
J. Kuhn, P. Vogt (eds.), *Smartphones as Mobile Minilabs in Physics*,
https://doi.org/10.1007/978-3-030-94044-7_61

based on alternating currents (AC), 50 Hz or 60 Hz, and, as a consequence, the created magnetic fields are difficult to measure. However, some railway electrification systems are powered by direct currents (DC) and thus the magnetic field produced by overhead lines (also known as catenaries) can occasionally be measured. In this work, we propose the analysis of the magnetic field produced by electric railways near Rome.

61.2   The Experiment

Modern smartphones usually possess built-in magnetometers that have been employed in several physics experiments [1–6]. The experiment proposed here was performed outdoors in a peaceful place near Rome. A pedestrian walking on an overpass above the tracks of the railways (shown in Fig. 61.1) registers the magnetic field with her/his smartphone. At that moment, no train was in the vicinity and no other vehicle was passing by. While the experimenter was walking at nearly constant speed ($\approx$ 1 m/s) the smartphone was held horizontally with the screen oriented upwards. This procedure is similar to the "flyby" on an air track proposed in Ref. [3]. Thanks to the phyphox app [7], the three components of the magnetic field were registered as the pedestrian walked over the bridge.

Fig. 61.1 The magnetic field measurement was taken by a pedestrian walking on an overpass above the tracks (left panel) shown in the satellite view (right panel). Axis x points in the direction of the currents through the tracks, opposite to the current I through the catenary (highlighted in yellow), while y is along the smartphone's path, and z points vertically upwards. From the NOAA website we obtained the horizontal component of Earth's magnetic field $B_{EH} = 24.6$ µT, the magnetic declination $\delta = 3.45°$, and the vertical component of Earth's magnetic field $B_{EZ} = -39.6$ µT. From the map, angle $\theta = 61.9°$

61.3 The Comparison

The magnetic field is the sum of the contributions of the magnetic fields of the right track (RT), left track (LT), catenary (C), and Earth (E), $B = B_{RT} + B_{LT} + B_C + B_E$, with the currents and distances sketched in Fig. 61.2. According to the train company, the power of the engines is 2.2 MW and the voltage of the lines is 3 kV (DC), resulting in an intensity of approximately $I = 733$ A, though this value can vary depending on the number of convoys nearby and their accelerations. We also assume a current in one direction through the catenary and the opposite, returning, current uniformly distributed in the two tracks. The orientation of the tracks, the path, and the Earth's magnetic field were obtained from the satellite view, and the reference frame is chosen to be the same of the smartphone, with the x-axis parallel to the tracks, the y-axis perpendicular to them and parallel to the overpass, and the z-axis vertical.

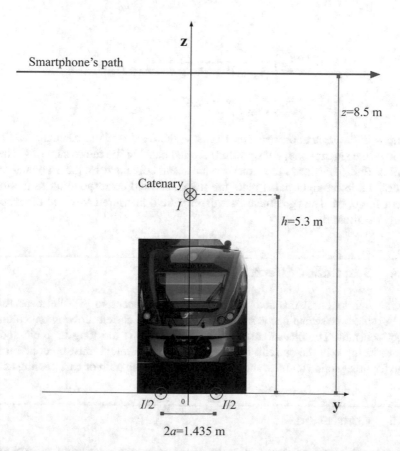

Fig. 61.2 Layout of the currents and the smartphone's path. The standard gauge railways and height of the catenary were taken from the website of the company while the height of the bridge was estimated from the pictures

The contributions produced by the currents are given by the Biot-Savart law for an infinite line

$$\boldsymbol{B} = \frac{\mu_0 i (-z)}{2\pi r^2} \widehat{y} + \frac{\mu_0 i (y)}{2\pi r^2} \widehat{z},$$

while from NOAA's geomagnetic calculator for that location, the components of Earth's magnetic field were obtained (Fig. 61.1). Then, the components along the axis shown in Figs. 61.1 and 61.2 can be written as

$$B_y = \frac{\mu_0}{2\pi} \left(\frac{I}{2}\right) \frac{(-z)}{(y-a)^2 + z^2} + \frac{\mu_0}{2\pi} \left(\frac{I}{2}\right) \frac{(-z)}{(y-a)^2 + z^2} +$$
$$\frac{\mu_0}{2\pi} (-I) \frac{(h-z)}{y^2 + (z-h)^2} + B_{EH} \cos(\theta - \delta)$$

and

$$B_z = \frac{\mu_0}{2\pi} \left(\frac{I}{2}\right) \frac{(y-a)}{(y-a)^2 + z^2} + \frac{\mu_0}{2\pi} \left(\frac{I}{2}\right) \frac{(y+a)}{(y+a)^2 + z^2} +$$
$$\frac{\mu_0}{2\pi} (-I) \frac{(y)}{y^2 + (z-h)^2} + B_{EZ},$$

where a is the separation between tracks, h the height of the catenary, and z the height of the smartphone, whose values are indicated in the figure captions. The left panel of Fig. 61.3 shows the experimental results obtained by the walking experimenter, while the right panel plots the magnetic field corresponding to the model described above. The agreement between the field measurements and the model is clearly manifested.

61.4 Systematic Effects

Despite smartphones' magnetometers not being so accurate [8], the measurement can be quite precise and needs some precautions. The current flowing on a train line is not constant. The measurement, then, must not last too long to avoid spotting sudden changes in the current. Moreover, the measurement must be done far from large ferromagnetic volumes, such as the wagons of the train or cars passing nearby.

61.5 Conclusion

The magnetic field produced by railways provides the possibility to explore the electromagnetic fields that surround us. Measurements can be successfully compared with estimations based on reasonable data and information available on the internet.

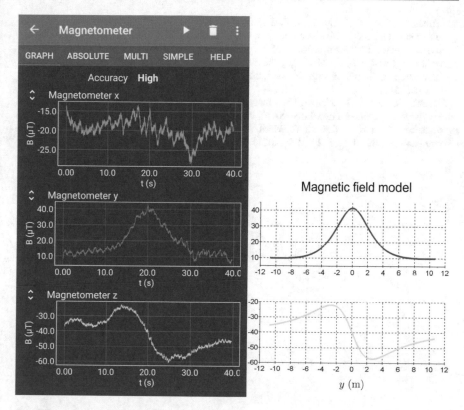

Fig. 61.3 Comparison between the measurements (phyphox screenshot on the left) and the model calculation (on the right)

This experiment encourages students to go outdoors and experiment using everyday tools.

References

1. Silva, N.: Magnetic field sensor. Phys. Teach. **50**, 372 (2012)
2. Septianto, R.D., Suhendra, D., Iskandar, F.: Utilisation of the magnetic sensor in a smartphone for facile magnetostatics experiment: Magnetic field due to electrical current in straight and loop wires. Phys. Educ. **52**(1), 015015 (2017)
3. Monteiro, M., Stari, C., Cabeza, C., Martí, A.C.: Magnetic field 'flyby' measurement using a smartphone's magnetometer and accelerometer simultaneously. Phys. Teach. **55**, 580 (2017)
4. Lara, V.O.M., Amaral, D.F., Faria, D., Vieira, L.P.: Demonstrations of magnetic phenomena: measuring the air permeability using tablets. Phys. Educ. **49**(6), 658 (2014)

5. Arribas, E., Escobar, I., Suarez, C.P., Najera, A., Beléndez, A.: Measurement of the magnetic field of small magnets with a smartphone: a very economical laboratory practice for introductory physics courses. Eur. J. Phys. **36**(6), 065002 (2015)
6. Arabasi, S., Al-Taani, H.: Measuring the Earth's magnetic field dip angle using a smartphone-aided setup: a simple experiment for introductory physics laboratories. Eur. J. Phys. **38**(2), 025201 (2017)
7. Staacks, S., Hütz, S., Heinke, H., Stampfer, C.: Advanced tools for smartphone-based experiments: phyphox. Phys. Educ. **53**(4), 045009 (2018)
8. Monteiro, M., Stari, C., Cabeza, C., Martí, A.C.: Using mobile-device sensors to teach students error analysis. Am. J. Phys. **89**, 477 (2021)

Magnetic Field 'Flyby' Measurement Using a Smartphone's Magnetometer and Accelerometer Simultaneously

62

Martín Monteiro, Cecilia Stari, Cecilia Cabeza, and Arturo C. Marti

The spatial dependence of magnetic fields in simple configurations is a common topic in introductory electromagnetism lessons, both in high school and in university courses. In typical experiments, magnetic fields and distances are obtained taking point-by-point values using a Hall sensor and a ruler, respectively. Here, we show how to take advantage of the smartphone capabilities to get simultaneous measures with the built-in accelerometer and magnetometer and to obtain the spatial dependence of magnetic fields. We consider a simple setup consisting of a smartphone mounted on a track whose direction coincides with the axis of a coil. While the smartphone is moving on the track, both the magnetic field and the distance from the center of the coil (integrated numerically from the acceleration values) are simultaneously obtained. This methodology can easily be extended to more complicated setups.

62.1 Simultaneous Use of Several Smartphone Sensors

Recently, the increasing availability and capabilities, and the decreasing cost, have contributed to the expansion of "smartphone physics." Indeed, smartphone sensors, as accelerometer, gyroscope, and magnetometer among others, have been successfully employed in diverse physics experiments ranging from mechanics to modern physics (see, for example, the iPhysicsLabs-column or the different chapters in this book). One relevant aspect that has received little attention is the fact that smartphones allow one to obtain simultaneous measures from several sensors. In

M. Monteiro (✉)
Universidad ORT Uruguay, Montevideo, Uruguay
e-mail: monteiro@ort.edu.uy

C. Stari · C. Cabeza · A. C. Marti
Universidad de la República, Montevideo, Uruguay
e-mail: cstari@fing.edu.uy; cecilia@fisica.edu.uy; marti@fisica.edu.uy

previous works, the simultaneous use of the accelerometer and the gyroscope has been proposed (Chaps. 18 and 39) [1–5]. More recently, the luminosity sensor has been employed together with the orientation sensor to experiment with polarized light [6]. In this work, a simple experience that combines the use of the smartphone magnetometer and the accelerometer is proposed. The magnetic field generated by a current in a coil is measured with a smartphone located over a cart on a track whose orientation coincides with the axis of the coil. While the smartphone is moving on the track, its position is readily obtained integrating twice the acceleration values obtained from the accelerometer. In this way, with a simple data processing, the magnetic field as a function of the position is obtained and as a by-product also the permeability. These results can be compared with the predictions of the Biot-Savart law.

62.2 Experiment

Until now a few smartphone-based experiments focusing on electromagnetism have been proposed in the literature (see for example Refs. [7–9]). In general, in these experiments the value of the magnetic field is obtained point by point and the distance is measured using a ruler like in traditional approaches. Here, we focus on the axial component of the magnetic field B along the axis of a thin coil carrying a current I. According to the Biot Savart law:

$$B = \frac{\mu_0 N I R^2}{2\left(R^2 + y^2\right)^{3/2}},$$

where R is the mean radius of the coil, μ_0 is the permeability of free space, N is the number of turns, and y is the distance to the center of the coil. The experimental setup, shown in Fig. 62.1, consists of a smartphone mounted on a track placed on the axis of a coil aligned with the track. The coil is made by $N = 200$ turns and its mean radius is $R = 10.3(2)$ cm. The smartphone is mounted on the track using an aluminum support to avoid magnetic interference. A DC power supply is used to create an electric current in the coil. The current intensity must be chosen so the maximum magnetic field that occurs at the center of the coil does not saturate the sensor. In the present experiment $I = 1.83(1)$ A.

The smartphone, a Nexus 5, contains a three-axis built-in accelerometer and magnetometer. First of all, we obtain the exact position of the magnetic sensor within the smartphone [9]. The position of the smartphone is chosen such that the y-axis coincides with the track and that the magnetic sensor is located on the axis of the coil. Second, it is necessary to consider that the sensor measures the magnetic field generated by the coil and also additional contributions by the terrestrial magnetic field and other nearby magnetic objects. To obtain the magnetic field produced only by the coil, the background contribution, obtained with zero applied current, must be subtracted. In this experiment the background magnetic field was

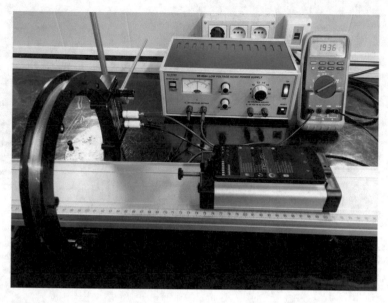

Fig. 62.1 Experimental setup

72.9 μT. The uncertainties in the values of the magnetic field and the acceleration obtained with the sensors were determined considering standard deviation in a measurement (0.6 μT and 0.02 m/s², respectively). To reduce the errors in the numerical integration, the sampling interval was chosen as the maximum rate of the accelerometer ($\Delta t = 0.005$ s).

The smartphone, initially at rest on the track, is gently propelled (with an acceleration of the order of 1.5 m/s² during a brief time interval) and continues following an almost uniform motion. The app Androsensor is used to record the y-components of both the acceleration and the magnetic field. The position of the smartphone along the track at the times at which the sensors measured a specific magnetic field value can be obtained integrating the acceleration values. In this case, the Euler method might be easily implemented in a spreadsheet. Concretely, given the initial conditions at time t_0, $x_0 = -0.30$ m and $v_0 = 0$, it is possible to iterate obtaining the variables at time t_{i+1} as a function of the variables at time t_i according to

$$v_{i+1} = v_i + a_i \Delta t$$
$$x_{i+1} = x_i + v_i \Delta t.$$

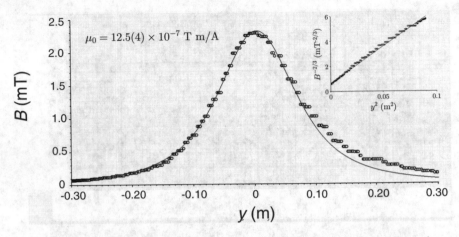

Fig. 62.2 Magnetic field as a function of the axial distance. The symbols represent the experimental results while the solid line corresponds to the Biot-Savart law. The first half of the path (−0.30 m, 0.0 m), where the numerical integration is more accurate, was used to build the linear fit displayed in the inset

62.3 Results and Analysis

In Fig. 62.2, the magnetic field as a function of the distance calculated through the numerical integration of the acceleration is displayed. The symbols correspond to the experimental measures while the theoretical model provided by the Biot-Savart law is presented by the red line. Due to the accumulation of error in the numerical integration, the agreement between the theoretical model and the experimental results is worse when the displacement is larger. The experimental value of the permeability is obtained from the slope, s, of the linear plot shown in the inset.

$$s = \left(\frac{\mu_0 NIR^2}{2} \right)^{-2/3}.$$

The experimental value, $\mu_0 = 12.5(4) \cdot 10^{-7}$ Tm/A, displays great concordance with the accepted value.

To sum up, we presented a simple and precise way to analyze the spatial dependence of magnetic fields using only a smartphone provided with acceleration and magnetic sensors. This approach can easily be extended to other experimental setups.

References

1. Monteiro, M., Cabeza, C., Marti, A.C.: Rotational energy in a physical pendulum. Phys. Teach. **52**, 561 (2014)

2. Monteiro, M., Cabeza, C., Marti, A.C., Vogt, P., Kuhn, J.: Angular velocity and centripetal acceleration relationship. Phys. Teach. **52**, 312–313 (2014)
3. Monteiro, M., Cabeza, C., Marti, A.C.: Exploring phase space using smartphone acceleration and rotation sensors simultaneously. Eur. J. Phys. **35**(4), 045013 (2014)
4. Vieyra, R., Vieyra, C., Jeanjacquot, P., Marti, A., Monteiro, M.: Five challenges that use mobile devices to collect and analyze data in physics. Sci. Teach. **82**(9), 32 (2015)
5. Monteiro, M., Cabeza, C., Marti, A.C.: Acceleration measurements using smartphone sensors: dealing with the equivalence principle. Revista Brasileira de Ensino de Fisica. **37**(1), 1303 (2015)
6. Monteiro, M., Stari, C., Cabeza, C., Marti, A.C.: The polarization of light and Malus' law using smartphones. Phys. Teach. **55**, 264 (2017)
7. Silva, N.: Magnetic field sensor. Phys. Teach. **50**, 372 (2012)
8. Lara, V., Amaral, D., Faria, D., Vieira, L.: Demonstrations of magnetic phenomena: measuring the air permeability using tablets. Phys. Educ. **49**(6), 658 (2014)
9. Arribas, E., Escobar, I., Suarez, C.P., Najera, A., Beléndez, A.: Measurement of the magnetic field of small magnets with a smartphone: a very economical laboratory practice for introductory physics courses. Eur. J. Phys. **36**(6), 065002 (2015)

Making the Invisible Visible: Visualization of the Connection Between Magnetic Field, Electric Current, and Lorentz Force with the Help of Augmented Reality

63

Anna Donhauser, Stefan Küchemann, Jochen Kuhn, Martina Rau, Sarah Malone, Peter Edelsbrunner, and Andreas Lichtenberger

When introducing electromagnetism in schools, one specific experiment is inevitable: the force on a current-carrying conductor. Predicting the direction of the Lorentz force, the orientation of the magnetic field, and the direction of the electric current often causes difficulties for students. Here we present visual concept-relevant augmentations of the experiment that use the Microsoft HoloLens, which intends to counteract common students' misconceptions by taking relevant principles of educational psychology into account.

63.1 Theoretical Background

Maxwell's equations [1] can completely describe the electrodynamics based on interaction of fields. Gauss's law of a magnetic field B is Maxwell's second equation for the magnetic flux:

A. Donhauser (✉) · S. Küchemann · J. Kuhn
Ludwig-Maximilians-Universität München (LMU Munich), Faculty of Physics, Chair of Physics Education, Munich, Germany
e-mail: a.donhauser@physik.uni-muenchen.de; jochen.kuhn@lmu.de

M. Rau
Department of Educational Psychology and Department of Computer Science, University of Wisconsin, Madison, WI, USA

S. Malone
Department of Education, Saarland University, Saarbrücken, Germany

P. Edelsbrunner
ETH Zürich, Department of Humanities, Social and Political Sciences/Chair for Learning and Instruction, Zürich, Switzerland

A. Lichtenberger
ETH Zürich, Department of Physics/Solid-State Dynamics and Education, Zürich, Switzerland

$$\oint_s \boldsymbol{B} dA = 0. \tag{63.1}$$

It can be clearly interpreted as the fact that there are no magnetic monopoles and that magnetic field lines always form closed loops.

Maxwell's fourth equation is formulated in the sense that an electric current I or a changing electric flux generates a circulating magnetic field:

$$\oint_l \boldsymbol{B} \cdot \mathrm{d}s = \mu_0 I + \varepsilon_0 \mu_0 \frac{\mathrm{d}}{\mathrm{d}t} \int_s \boldsymbol{E} \cdot \mathrm{d}\boldsymbol{A}. \tag{63.2}$$

Complementing Maxwell's equations with the Lorentz force can describe all (classical) electrodynamic phenomena. This Lorentz force $\boldsymbol{F}_{\mathrm{L}}$ results from the vector product of the charge Q, which passes through a magnetic field $\boldsymbol{B}$ with velocity $\boldsymbol{v}$:

$$F_{\mathrm{L}} = Q\boldsymbol{v} \cdot \boldsymbol{B}.$$

For a current-carrying wire section of length dl, this equation can be transformed using the relation

$$Q\boldsymbol{v} = Q\frac{\mathrm{d}l}{\mathrm{d}t} = \frac{\mathrm{d}Q}{\mathrm{d}t}\mathrm{d}l = I \cdot \mathrm{d}l$$

in the equation of the Lorentz force, which acts on a conductor with current I:

$$F_{\mathrm{L}} = I \cdot \mathrm{d}l \times \boldsymbol{B}.$$

63.2 Experimental Setup and Procedure

In the experiment of a current-currying conductor, students can directly explore the phenomena that are formally described by Maxwell's equations. Figure 63.1 shows the usual setup with a freely suspended current-carrying conductor placed in the magnetic field of a horseshoe magnet. A power supply generates a current of 10 A in the conductor swing.

Two problems arise for our students in this experiment. First, field effects on current-currying conductors are invisible. Second, split attention effects following from the discontinuity of relevant information impairs learning, because related additional representations such as vector representation, formulas, and magnetic field lines are presented at a spatial or temporal distance from the experimental setup (e.g., on the blackboard after experimenting). As a result, students either completely ignore invisible fields, misinterpret models of the invisible fields, or they incorrectly predict directions of interaction and movement. Therefore, the representations and their directions of interaction might help students to develop physically correct mental concepts and problem-solving strategies.

Fig. 63.1 Experimental
setup for the demonstration of
the Lorentz force

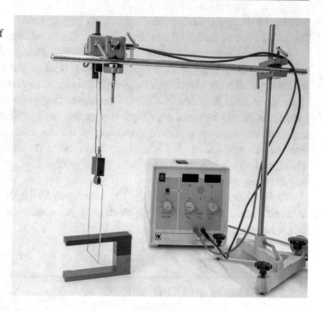

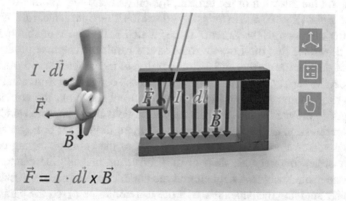

$$\vec{F} = I \cdot d\hat{l} \times \vec{B}$$

Fig. 63.2 Visualization of the magnetic field and the right-hand rule using the HoloLens. Option-
ally, three learning tools can be displayed: the formula of the Lorentz force, directional indications
with the right hand, and the vector triad

We address this with augmented reality (AR), as shown in Fig. 63.2. Based on
prior work (Chaps. 53 and 57) [2–6], we use a HoloLens [7] to combine a virtual
visualization with the physical experiment. Using these mixed-reality glasses, we
create an interactive learning environment that follows well-established multimedia
learning design principles [8]. This not only makes static invisibilities visible, but
also allows students to directly experience physical relationships and processes
because they see in real time how the parameters they modified dynamically
influence other variables, processes, and fields. Thus, we obtain spatial and temporal

contiguity between reality and its virtual representations to reduce a spatial and temporal split of students' attention during the experiment.

As Fig. 63.2 shows, the HoloLens virtually visualizes the model of the magnetic field lines directly around the magnet. The model of the magnetic field combines a continuum with the directional magnetic field lines to prevent the misconception that there is no field between the lines. Corresponding vectors, field lines, and physical quantities in the formula are displayed in the same color. This color-coding helps students relate corresponding representations to each other. In addition to the presented experiment, the magnetic field around the conductor swing will be visualized. This allows students to observe the superposition and interaction of several magnetic fields.

To study the dependence of the Lorentz force F_L on the parameters of the electric current I and the conductor length dl, we subsequently manipulate each variable while keeping the others constant.

63.3   Results and Outlook

Depending on the direction of the current, the current-carrying conductor moves to the left or right (Fig. 63.2). In physics education the right-hand rule is used to illustrate the direction of the Lorentz force, which is the cause of this movement. This rule illustrates that the Lorentz force is the result of the cross product of the direction of the electric current and the orientation of the magnetic field. As there are country-specific variations of this rule, we follow the convention shown in Fig. 63.2: The right-hand thumb points in the technical current direction, the corresponding index finger is oriented parallel to the magnetic field lines, and the middle finger predicts the direction of the Lorentz force. All three fingers are orthogonal to each other like a vector triad. Whereas the movement of the current-carrying conductor becomes visible, the magnetic field of the horseshoe magnet, the electric current flow, and the related magnetic field around the conductor swing remain invisible. By subsequently studying the dependence of the Lorentz force F_L on the parameters of the electric current I and the conductor length dl, students can learn that the Lorentz force is proportional to the electric current I and the conductor length dl.

While the visualization of invisible, abstract variables should lead to better learning, the multitude of representations that can possibly be integrated may also increase cognitive load. Therefore, future work should investigate which forms of representations maximize students' learning and how many representations are optimal to maximally support students without cognitively overloading them. This important as research shows that effects on learning in educational lab work instruction are not consistent [6, 9, 10].

References

1. Demtröder, W.: Electrodynamics and Optics, p. 110. Springer International Publishing, Cham (2019)
2. Kuhn, J., Lukowicz, P., Hirth, M., Poxrucker, A., Weppner, J., Younas, J.: gPhysics – using smart glasses for head-centered, context-aware learning in physics experiments. IEEE Trans. Learn. Technol. **9**(4), 304 (2016)
3. Strzys, M.P., Kapp, S., Thees, M., Kuhn, J., Lukowicz, P., Knierim, P., Schmidt, A.: Augmenting the thermal flux experiment: a mixed reality approach with the HoloLens. Phys. Teach. **55**, 376 (2017)
4. Strzys, M.P., Kapp, S., Thees, M., Klein, P., Lukowicz, P., Knierim, P., Schmidt, A., Kuhn, J.: Physics holo.lab learning experience: using smartglasses for augmented reality labwork to foster the concepts of heat conduction. Eur. J. Phys. **39**(3), 035703 (2018)
5. Kapp, S., Thees, M., Strzys, M.P., Beil, F., Kuhn, J., Amiraslanov, O., Javaheri, H., Lukowicz, P., Lauer, F., Rheinländer, C., Wehn, N.: Augmenting Kirchhoff's laws: using augmented reality and smartglasses to enhance conceptual electrical experiments for high school students. Phys. Teach. **57**, 52 (2019)
6. Thees, M., Kapp, S., Strzys, M.P., Beil, F., Lukowicz, P., Kuhn, J.: Effects of augmented reality on learning and cognitive load in university physics laboratory courses. Comp. Hum. Behav. **108**, 106316 (2020)
7. https://www.microsoft.com/en-us/hololens (This is an independent publication and is neither affiliated with, nor authorized, sponsored, or approved by, Microsoft Corp.)
8. Chien, K.P., Tsai, C.Y., Chen, H.L., Chang, W.H., Chen, S.: Learning differences and eye fixation patterns in virtual and physical science laboratories. Comput. Educ. **82**, 191–201 (2015)
9. Altmeyer, K., Kapp, S., Thees, M., Malone, S., Kuhn, J., Brünken, R.: Augmented reality to foster conceptual knowledge acquisition in STEM laboratory courses – theoretical derivations and empirical findings. Brit. J Educ. Technol. **51**(3), 611–628 (2020)
10. Thees, M., Altmeyer, K., Kapp, S., Rexigel, E., Beil, F., Malone, S., Brünken, R. & Kuhn, J. (2022). Augmented Reality for Presenting Real-Time Data During Students Laboratory Work: Comparing Smartglasses with a Separate Display. *Front. Psych. 13* (2022), 804742

Augmenting the Fine Beam Tube: From Hybrid Measurements to Magnetic Field Visualization

64

Oliver Bodensiek, Dörte Sonntag, Nils Wendorff,
Georgia Albuquerque, and Marcus Magnor

Since the emergence of augmented reality (AR), it has been a constant subject of educational research, as it can improve conceptual understanding and generally promote learning [1]. In addition, a motivational effect and improved interaction and collaboration through AR were observed [2]. Recently, AR technologies have taken a major leap forward in development, such that head-mounted devices or smartglasses in particular are now finding their first applications in STEM education, especially in experiments (Chaps. 53 and 57) [3–6]. In line with these developments, we here present an AR experiment in electrodynamics for undergraduate laboratory courses in physics using real-time physical data from and virtual tools on mobile devices to both analyze and visualize physical phenomena.

64.1 Theoretical Background

In order to determine the electron charge-to-mass ratio $-e/m_e$, a fine beam tube is typically used in educational settings. Its main part is an electron gun that generates electrons by thermal glow emission, accelerates them due to a voltage U_{acc} between anode and cathode, and bundles them into a focused beam. The electron gun is embedded into an evacuated glass sphere back-filled with hydrogen or helium at low pressure. Hence, the electron beam becomes visible due to impact ionization. This fine beam tube is mounted on a stand right in the middle of a Helmholtz coil pair, where a coil current I_{coil} generates an almost homogenous magnetic field. Provided I_{coil} is large enough compared to U_{acc}, the resulting Lorentz force deflects the

O. Bodensiek (✉) · D. Sonntag
Institute for Science Education Research, Physics Group, Technische Universität Braunschweig,
Braunschweig, Germany
e-mail: o.bodensiek@tu-braunschweig.de

N. Wendorff · G. Albuquerque · M. Magnor
Computer Graphics Lab, Technische Universität Braunschweig, Braunschweig, Germany

J. Kuhn, P. Vogt (eds.), *Smartphones as Mobile Minilabs in Physics*,
https://doi.org/10.1007/978-3-030-94044-7_64

electrons onto a circular path within the glass sphere. The Lorentz force acting on an electron with velocity v_e in a magnetic field B is given by

$$F = -e(v_e \times B).$$

For an ideal homogenous magnetic field and the electron velocity being perpendicular to the magnetic field, the Lorentz force acts as radial force, that is,

$$m_e \cdot \frac{v_e^2}{r} = e \cdot v_e \cdot B$$

according to amount and with r representing the radius of the circular electron path. Using the relations between v_e and U_{acc}, respectively, between B and I_{coil}, the essential proportionality in this experiment is given by

$$\frac{e}{m_e} \propto \frac{U_{acc}}{r^2 \cdot I_{coil}^2}.$$

64.2   Experimental Setup

In the experiment students use the Microsoft HoloLens [7] as AR smartglasses both to record measurement data and to study the physics of charged particles in magnetic fields in a hybrid, i.e., digitally enhanced lab environment.

U_{acc} and I_{coil} are measured with multimeters and gathered by a USB-connected single board computer (Fig. 64.1), which in turn establishes a wireless data link to

Fig. 64.1 Experimental setup of the fine beam tube with Helmholtz coils, power supplies, and multimeters connected via USB with a single board computer (SBC)

Fig. 64.2 View through the smartglasses while using the virtual ruler

the HoloLens. On the HoloLens's semi-transparent display, the real-time measurement data are presented as numerical values in real time (Fig. 64.2). In order to determine the radius of the electron beam, we have added a virtual ruler that can by gesture control both be moved in depth to the plane of the electron beam and adjusted to the diameter of the circular beam path (Fig. 64.2).

All numerical values that are needed to calculate the charge-to-mass ratio are shown on the display of the smartglasses. By "air tapping" a single record button (out of the field of view in Fig. 64.2), all three values are automatically added to a CSV file that can be analyzed after the experiment. Additional measurements are recorded either for different voltages and coil currents measuring the altered diameter again, or by adjusting U_{acc} and I_{coil} such that the diameter keeps constant. Especially in the latter case, a measurement series can be done both rather quickly and by a single student, if wanted or necessary.

In addition to the hybrid measurement possibility, we implemented a visualization of the magnetic field (Fig. 64.3) according to the current experimental parameters. The user can also overlay the corresponding theoretically predicted electron beam. All field and beam data are calculated in advance by the finite-element method for a relevant parameter space of U_{acc} and I_{coil} with a sufficiently narrow parameter grid. In between these points on the grid, the AR application interpolates the pre-calculated field and beam data linearly and visualizes it on the smartglasses accordingly. Moreover, the relevant formulas for Lorentz force, field strength as function of I_{coil}, and electron velocity as function of U_{acc} can be enabled as overlay to the experiment (not shown here). The parameter(s) currently being changed in the experiment are then highlighted by color in the formulas.

Fig. 64.3 Visualization of the magnetic field as vector plot. In addition, the theoretically predicted electron beam is augmented

64.3 Experimental Results

We recorded a series of $N = 18$ measurements with constant radius but different acceleration voltages and coil currents in the AR environment. As a mean value we obtained $-e/m_e = -(1.76 \pm 0.01) \cdot 10^{11}$ C·kg^{-1}, which is remarkably close to the CODATA value $-1.758820024(11) \cdot 10^{11}$ C·kg^{-1} [8]. In several manual measurement series, visually reading off the values, we best reached an accuracy of only about 3%. We relate the improved accuracy in the AR environment to the following two factors: on the one hand, even the stabilized power and voltage supplies we used vary over time. As an effect, one or both values read off may differ a little from the ones immediately after calibration to the constant radius. This error source is reduced in our AR measurement, where both values are recorded simultaneously right after calibration. On the other hand, visually reading off the diameter includes a parallax error, as the measuring device lies approximately 9 cm in front of the electron beam. This error can be reduced by using a mirror on the backside of the glass sphere, but it is still not as accurate as the virtual ruler in our AR environment, which can be placed directly in the plane of the electron beam in the glass sphere.

The only drawback we experienced in using the AR environment for this specific experiment is the need to adjust the light conditions so that the weakly glowing electron beam is clearly visible through the darkening HoloLens and one can still see enough to operate the power supplies.

64.4 Conclusion

We implemented an augmented lab experiment for the fine beam tube using AR smartglasses. All measurements in order to determine the specific charge can be recorded digitally in the AR environment. With this AR-based approach, we observe several advantages. First, the measured values seem to be more accurate compared to reading them off visually. Second, acquisition of measurement values is easy and quick and can easily be done alone. Finally, and probably most important, the additional field visualization coupled to real-time data provides an immediate feedback to the students' experimental actions. In combination with corresponding mathematical formulas of a theoretical description in a single hybrid learning environment, we expect this to foster understanding relationships between theory and experiment as found in comparable AR experiments [6] since it provides high temporal and spatial contiguity, thereby avoiding a split-attention effect [9].

References

1. Cheng, K.-H., Tsai, C.-C.: Affordances of augmented reality in science learning: suggestions for future research. J. Sci. Educ. Technol. **22**, 449–462 (2013)
2. Bacca, J., Baldiris, S., Fabregat, R., Graf, S., Kinshuk: Augmented reality trends in education: a systematic review of research and applications. Educ. Technol. Soc. **17**, 133–149 (2014)
3. Strzys, M.P., Kapp, S., Thees, M., Kuhn, J., Lukowicz, P., Knierim, P., Schmidt, A.: Augmenting the thermal flux experiment: a mixed reality approach with the HoloLens. Phys. Teach. **55**, 376–377 (2017)
4. Kapp, S., Thees, M., Strzys, M.P., Beil, F., Kuhn, J., Amiraslanov, O., Javaheri, H., Lukowicz, P., Lauer, F., Rheinlander, C., Wehn, N.: Augmenting Kirchhoff's laws: using augmented reality and smartglasses to enhance conceptual electrical experiments for high school students. Phys. Teach. **57**, 52–53 (2019)
5. Kuhn, J., Lukowicz, P., Hirth, M., Poxrucker, A., Weppner, J., Younas, J.: gPhysics—using smart glasses for head-centered, context-aware learning in physics experiments. IEEE Trans. Learn. Technol. **9**, 304–317 (2016)
6. Strzys, M.P., Kapp, S., Thees, M., Klein, P., Lukowicz, P., Knierim, P., Schmidt, A., Kuhn, J.: Physics holo.lab learning experience: Using smartglasses for augmented reality labwork to foster the concepts of heat conduction. Eur. J. Phys. **39**, 35703 (2018)
7. Microsoft HoloLens. https://www.microsoft.com/en-us/hololens
8. Mohr, P.J., Newell, D.B., Taylor, B.N.: CODATA recommended values of the fundamental physical constants: 2014. Rev. Mod. Phys. **88**, 337 (2016)
9. Mayer, R. (ed.): The Cambridge Handbook of Multimedia Learning. Cambridge University Press, Cambridge (2005)

Learning the Lens Equation Using Water and Smartphones/Tablets

65

Jack Freeland, Venkata Rao Krishnamurthi, and Yong Wang

Due to their prevalence and convenience, smartphones and tablets have been increasingly useful as tools to carry out new experiments in STEM education, including high school and university physics such as mechanics, acoustics, and optics (e.g. Chaps. 6, 24, 29 and 44) [1–8]. In this chapter we present a simple experiment with water and smartphones/tablets. We demonstrate how smartphones and tablets can be used for learning and verifying the lens equation and lensmaker's equation to help students understand concepts in ray optics.

For a convex (converging) lens with a focal length of f, if the distances from the lens to an object and its image are r_o and r_i (Fig. 65.1a), respectively, the lens equation is

$$\frac{1}{r_o} + \frac{1}{r_i} = \frac{1}{f}. \tag{65.1}$$

Given the radii of the lens surfaces, R_1 and R_2, the focal length can be obtained from the lensmaker's equation [9]

$$\frac{1}{f} = (n-1)\left[\frac{1}{R_1} + \frac{1}{R_2} - \frac{(n-1)w}{nR_1R_2}\right], \tag{65.2}$$

where n is the refractive index of the lens material, and $w > 0$ is the thickness of the lens. In this experiment, we will create a thick convex (converging) lens by putting a

J. Freeland · V. R. Krishnamurthi
Department of Physics, University of Arkansas, Fayetteville, AR, USA
e-mail: jfreeland@mednet.ucla.edu

Y. Wang (✉)
Department of Physics, Materials Science and Engineering Program, and Cell and Molecular Biology Program, University of Arkansas, Fayetteville, AR, USA
e-mail: yongwang@uark.edu

© The Author(s), under exclusive license to Springer Nature Switzerland AG 2022
J. Kuhn, P. Vogt (eds.), *Smartphones as Mobile Minilabs in Physics*,
https://doi.org/10.1007/978-3-030-94044-7_65

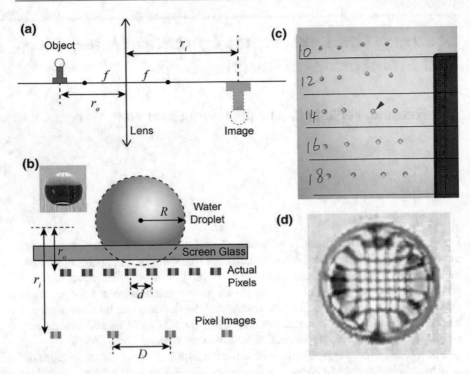

Fig. 65.1 Theory and experiment. (**a**) Lens equation. (**b**) Magnification of pixels of a screen by a water droplet. Inset: side view of a water droplet. (**c**) An example of pictures of water droplets of different volumes on an iPad screen. (**d**) Zoom-in view of a droplet in panel (**c**) indicated by the red arrow

water droplet on the screen of a smartphone or tablet (Fig. 65.1b). As the screen side of the water droplet is flat, we have $R_1 = \infty$. As the water droplet can be approximated as part of a sphere, R_2 is approximately the radius of the sphere R. Taking into account that the refractive index of water is $n = 1.333$, the lens equation [Eq. (65.1)] is simplified to

$$\frac{1}{r_o} + \frac{1}{r_i} = \frac{1}{f} = \frac{n-1}{R} = \frac{1}{3R}. \tag{65.3}$$

If the magnification is negative (i.e., $r_i/r_o < 0$) we can define and write

$$\beta = 1 - 1/M = r_o/(3R), \tag{65.4}$$

where $M = |r_i/r_o|$. Thus, it is expected that β varies linearly with $1/R$ if r_o is constant.

65.1   Laboratory Experiment

In the laboratory experiment, lenses made of water droplets were created and used to magnify the pixels of the screens of smartphones/tablets, and the lens equation was verified. Briefly, an iPad (6th generation, Model MR7K2LL/A) was placed (with its screen facing up) on a flat surface. Before placing water droplets on the iPad screen, students were advised to open an app (e.g., GoodNotes) to generate a uniform white background and draw several horizontal lines on the screen as guides (black lines in Fig. 65.1c). In addition, students were required to write down the desired volumes (in mL) of water droplets (the numbers—10, 12, 14, 16, and 18—in Fig. 65.1c). The guides and numbers help students to organize the water droplets and the data to be acquired. Furthermore, a ruler with markings was placed on the iPad close to the edge of the screen. The ruler is needed for estimating and calibrating the pixel size of pictures of water droplets in later analysis.

Water droplets of different volumes (10–18 μL) were added to the iPad screen using a 20 μL pipettor (VWR International LLC). Each volume was repeated four times (i.e., four droplets in a row as shown Fig. 65.1c). Due to the presence of the water droplet lenses, students observed that the actual pixels (d) of the iPad screen formed virtual images (D) (Fig. 65.1b, c). This observation was recorded as pictures by the camera of a smartphone (iPhone 7, Model # MN8G2LL/A). A closer look at single droplets in the acquired pictures clearly showed the pixel images of the iPad screen (Fig. 65.1d). The acquired pictures were then transferred to a computer and analyzed.

It is suggested that students take pictures of all the water droplets together (as shown in Fig. 65.1c), instead of separate pictures of individual water droplets. We found that the suggested procedure resulted in higher efficiency and, more importantly, better consistency. This is because the pixel size of the acquired pictures depends on the height of the smartphone, while a single picture including all the water droplets ensures that the height of the smartphone remains the same for all the water droplets in that picture.

65.2   Analysis and Results

Estimation of the Radius of Curvature of Water Lenses

To determine the radii of water droplets and magnification of the water lenses, we first quantified the pixel size of the captured pictures. Briefly, the picture was opened in ImageJ (an open-source software) [10] and a single color channel was extracted (ImageJ menu: Image → Color → Split → Channels). We then drew a line along the marks of the ruler (yellow line in Fig. 65.2a) and determined the intensity profile (ImageJ menu: Analyze → Plot Profile), from which a periodic pattern was observed (Fig. 65.2b). We next read out the periods (in the unit of pixels) from ImageJ [10], which showed a clear single-peak distribution centered at 26 pixels (Fig. 65.2c). As the periodic pattern corresponds to the mm marks on the ruler, one pixel in the

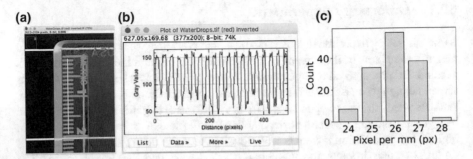

Fig. 65.2 Determining the pixel size of the captured picture. (**a**) Zoom in to the end of the ruler in the full picture. (**b**) Intensity profile along the yellow line in panel (**a**). (**c**) Distribution of the number of pixels between mm marks of the ruler

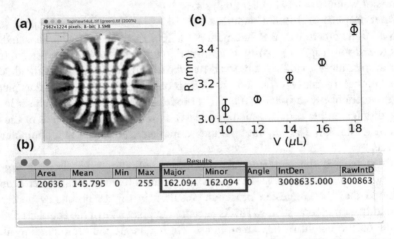

Fig. 65.3 Determining the size of water droplets. (**a**) Boundary of a droplet. (**b**) Measured major and minor axes (in the unit of pixels). (**c**) Dependence of the radii (R) of water droplets on their volume (V). Error bars stand for standard deviation from multiple measurements

captured pictures was then estimated at $\approx 1/26$ mm (i.e., the conversion factor was θ $\approx 1/26$ mm/px).

The radii of the water droplets were estimated using the oval selection tool and the Measure menu in ImageJ [10], and the determined conversion factor θ. Briefly, an oval selection was drawn to overlap with the boundary of the water droplet (Fig. 65.3a), followed by measuring the size of the oval (ImageJ menu: Analyze $\rightarrow$ Measure). From the major and minor axes (a and b, in the unit of pixels; Fig. 65.3b) and the conversion factor θ, the radius of the water droplet was estimated by $R = \theta\sqrt{ab}$ (in the unit of mm), which increased as the volume of the water droplets was higher (Fig. 65.3c).

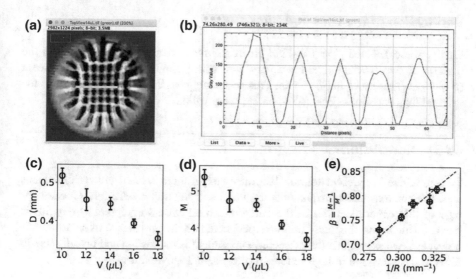

Fig. 65.4 Verification of lens equation. (**a**) Inverted picture of a droplet. (**b**) Intensity profile along the yellow line in panel (**a**). (**c**) Dependence of the magnified distance between screen pixels (D) on the volume of water droplets (V). (**d**) Dependence of the magnification factor (M) of water droplets on their volume (V). (**e**) Relation between $\beta = (M-1)/M$ and $1/R$. The red dashed line is the fitting with a line with zero-intercept. Error bars stand for standard deviation from multiple measurements

Determination of the Magnification of Water Lenses

The magnification of the water droplet lenses was determined from the magnified distance (D) and actual distance (d) between the screen pixels, $M = D/d$ (Fig. 65.1b). The actual distance d can be obtained from the specification of the iPad (pixels per inch or ppi = 264),[1] $d = 25.4/264$ mm $= 0.0962$ mm. To get the magnified distance (D), we first inverted the grayscale image (single color channel) of the water droplets and drew yellow lines horizontally or vertically (Fig. 65.4a). The intensity profile along the line was obtained (Fig. 65.4b), ImageJ menu: Analyze $\rightarrow$ Plot Profile). From the peak-to-peak distances L_{pp} (in the unit of pixels) read out from the intensity profile in ImageJ, we calculated the magnified distances of the iPad-screen pixels, $D = L_{pp} \cdot \theta$ (in the unit of mm), which gave the magnifications of the water lenses using $M = D/d$. Interestingly, it was observed that the magnified distances between screen pixels (D) and the magnification factor (M) were negatively correlated to the volume of the water droplets (Fig. 65.4c, d).

---

[1] iPad (6th generation)—technical specifications.

Verification of the Lens Equation

Lastly, we calculated $\beta = (M-1)/M$ from the magnifications and found that the relation of β vs. $1/R$ could be fitted very well with a single line with zero-intercept ($\beta = k \cdot 1/R$) as shown in Fig. 65.4e. This result suggests that Eq. (65.4) and thus the original lens equations [Eqs. (65.1, 65.3)] were valid.

65.3   Conclusions

To summarize, we present a simple laboratory experiment to learn and verify the lens equation and lensmaker's equation with water and smartphones/tablets. One advantage of this experiment lies in the fact that no additional equipment or optics is needed. Due to the simplicity of this experiment and the prevalence of smartphones nowadays, we expect that the presented experiment is interesting and broadly useful for physics education in high schools, colleges, and universities.

Acknowledgment This work was supported by the University of Arkansas; the iPad used in this work was provided by the Honors College at the University of Arkansas. We thank Angeli X. Wang, who observed the lens effects of water droplets during one of her Water Play Days, for inspiring and initiating this project.

References

1. Vogt, P., Kuhn, J.: Analyzing simple pendulum phenomena with a smartphone acceleration sensor. Phys. Teach. **50**, 439 (2012)
2. Vogt, P., Kuhn, J.: Analyzing free fall with a smartphone acceleration sensor. Phys. Teach. **50**, 182 (2012)
3. Chevrier, J., et al.: Teaching classical mechanics using smartphones. Phys. Teach. **51**, 376 (2013)
4. Goy, N.-A., et al.: Surface tension measurements with a smartphone. Phys. Teach. **55**, 498 (2017)
5. Hergemöller, T., Laumann, D.: Smartphone magnification attachment: Microscope or magnifying glass. Phys. Teach. **55**, 361 (2017)
6. Hirth, M., Kuhn, J., Müller, A.: Measurement of sound velocity made easy using harmonic resonant frequencies with everyday mobile technology. Phys. Teach. **53**, 120 (2015)
7. Thoms, L.-J., Colicchia, G., Girwidz, R.: Audiometric test with a smartphone. Phys. Teach. **56**, 478 (2018)
8. Shakur, A., Kraft, J.: Measurement of Coriolis acceleration with a smartphone. Phys. Teach. **54**, 288 (2016)
9. Hecht, E.: Optics. Pearson Education (2003)
10. Schneider, C.A., Rasband, W.S., Eliceiri, K.W.: NIH Image to Image J: 25 years of image analysis. Nat. Methods. **9**, 671 (2012)

Color Reproduction with a Smartphone

66

Lars-Jochen Thoms, Giuseppe Colicchia, and Raimund Girwidz

The world is full of colors. Most of the colors we see around us can be created on common digital displays simply by superposing light with three different wavelengths. However, no mixture of colors can produce a fully pure color identical to a spectral color. Using a smartphone, students can investigate the main features of primary color addition and understand how colors are made on digital displays.

66.1 Color Vision

Color perception with a high spectral resolution would need a large number of photoreceptors sensitive to narrow bands of light. Since an expansive number of photoreceptors on the retina would reduce the spatial resolution, eyes of living beings combine signals of a few photoreceptors, which are sensible on a large spectral range [1].

In our retina, there are three types of color receptors: long (L; red), medium (M; green), and short (S; blue) sensitive cones (Fig. 66.1) that owe their names to the wavelength of their peak sensitivities. The sensitivities of the cones overlap massively in the spectral range between 390 nm and 720 nm. Energy corresponding to the wavelength of light wave stimulates the cones, which send a stimulus to the brain. Specific combinations of stimuli from the three different cone receptor types produce an input from which the brain creates the perception of color.

The stimulus of one cone alone is not sufficient to identify the distribution of spectral intensities. However, each pulse of light radiation with a single (relative narrow band) wavelength between 400 and 700 nm stimulates two or three receptors, inducing the sensation of a specific spectral color. For example, light with frequency

L.-J. Thoms (✉) · G. Colicchia · R. Girwidz
Department of Physics/Chair of Physics Education, Ludwig-Maximilians-Universität München, Munich, Germany
e-mail: l.thoms@lmu.de

© The Author(s), under exclusive license to Springer Nature Switzerland AG 2022
J. Kuhn, P. Vogt (eds.), *Smartphones as Mobile Minilabs in Physics*,
https://doi.org/10.1007/978-3-030-94044-7_66

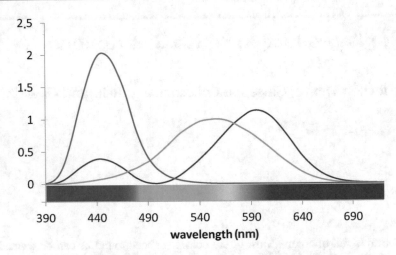

Fig. 66.1 The spectral sensitivity of the L, M, and S cones based on the CIE 1964 color matching functions. CIE is the *Commission Internationale de l'Eclairage* (International Commission on Illumination)

580 nm stimulates cones M and S equally and produces the sensation of the color yellow. However, light composed of both the two wavelengths 570 nm and 590 nm produces a stimulus that is almost equal to the sensation of the spectral color yellow. This is called "metamerism" and will be illustrated in the last experiment.

66.2 Additive Color Mixing

Like computer monitors and televisions, smartphones apply additive color mixing. These devices generate images using a tightly packed mosaic of red, green, and blue dots (Fig. 66.2) (RGB system). These dots are so small that our eyes cannot optically

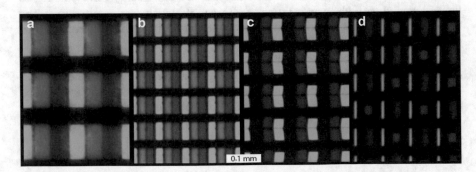

Fig. 66.2 Smartphone displays under a microscope. (**a**) iPhone 3GS (163 ppi), (**b**) iPhone 4 (326 ppi), (**c**) HTC 7 Mozart (252 ppi), (**d**) Samsung Galaxy S3 (306 ppi)

resolve them. Therefore, the light from the dots overlaps on the retina, creating a composite color impression.

Superposition with different intensities of these three primary colors can produce a wide range of colors. Per definition, no combination of two primary colors can produce a third primary color. The addition of two primary colors with identical intensities results in the secondary colors cyan, magenta, and yellow. Combining all three primary colors with even intensities leads to the color white.

66.3 Analyzing Colors of RGB Displays

Students can use any color mixer or any graphical software on a smartphone to identify the three primary colors of the RGB system: red, green, and blue. They simply draw vertical colored lines on a black screen, and examine them through inexpensive diffraction glasses[1] (Fig. 66.3).

Figure 66.4 shows red, blue, and green lines producing monochromatic diffraction images. In spite of that, a white line splits into the colors red, green, and blue. A yellow line shows a combination of green and red.

Each pixel in the display generates color by combining one red, one green, and one blue light dot. For example, when the red dot is set to zero, the light dot is turned off. When it is set to 255, the light dot is turned fully on. Any value between them sets the light dot to partial light emission.

Note that different devices display a particular RGB value in different ways, since the color elements (such as phosphors or dyes) vary from manufacturer to manufacturer. Thus, an RGB value does not describe the same color across devices without additional color management.

66.4 Two Kinds of Yellow

A pure spectral color stands for a single wavelength with a very narrow natural line width. Project the cone of light from a sodium vapor lamp as a spot on a white wall. Now take a photograph of the spot with a smartphone or a digital camera. Compare the picture with the spot on the wall. Both appear to be the same kind of yellow color. When you look through diffraction glasses at the spot on the wall, you will see three diffracted orders in pure yellow. However, when looking at the picture on the smartphone display, the higher orders split up in the three colors red, green, and blue.

As an alternative to expensive sodium vapor lamps, you can use a special candle bulb that contains LEDs and which is covered with a nearly monochromatic filter and view it directly[2] (Fig. 66.5).

---

[1] Diffraction glasses are available at, e.g., https://www.rainbowsymphony.com/
[2] For example, Paulmann LED 230 V, 0.6 W, E14, yellow.

Fig. 66.3 Diffraction glasses

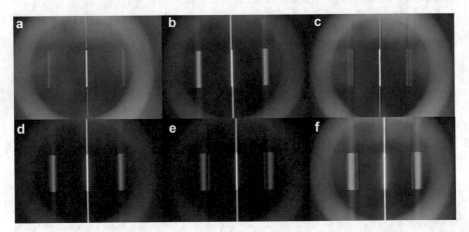

Fig. 66.4 Spectra of pure red (**a**), pure green (**b**), pure blue (**c**), yellow (**d**), and white with two different intensities (**e**, **f**)

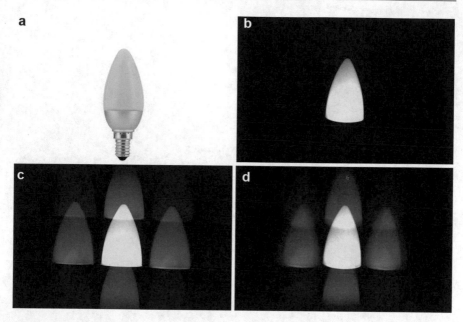

Fig. 66.5 (a) Candle bulb covered with a monochromatic yellow filter. (b) Image of the glowing bulb as seen with the naked eye and on a smartphone. (c) Photograph of the bulb through diffraction glasses, and (d) photograph of the smartphone display through diffraction glasses. The printed colors may appear different from those on screen, which in turn may appear different from those observed in the experiment

Reference

1. Osorio, D., Vorobyev, M.: Photoreceptor spectral sensitivities in terrestrial animals: Adaptations for luminance and colour vision. Proc. R. Soc. B. **272**, 1745–1752 (2005)

Diffraction Experiments with Infrared Remote Controls

67

Jochen Kuhn and Patrik Vogt

In this contribution we describe an experiment in which radiation emitted by an infrared remote control is passed through a diffraction grating. An image of the diffraction pattern is captured using a cell phone camera and then used to determine the wavelength of the radiation Corresponding ideas were previously published in [1–3].

The CCD chips used in digital cameras are also sensitive to electromagnetic waves in the near infrared. This property can be used to demonstrate interesting diffraction phenomena with simple apparatus. In addition to a cell phone with a camera function, the objects required for this experiment are an infrared remote control and a diffraction grating with a suitable line spacing (50 lines per millimeter, or so—low-cost transmission grating film works well, or even a CD used as a reflection grating).

67.1 Qualitative Experiments

The experimental setup is very simple: The user shines the remote control onto the camera lens while holding the grating directly in front of the lens.

The diffraction pattern can be photographed using the cell phone camera (Fig. 67.1) and exported to a computer to be printed and inserted into students' lab reports. In Fig. 67.2 the image on the left was produced by a transmission grating having 50 lines per millimeter, and the image on the right by a cross-grating film with 900 lines per millimeter.

J. Kuhn (✉)
Ludwig-Maximilians-Universität München (LMU Munich), Faculty of Physics, Chair of Physics Education, Munich, Germany
e-mail: jochen.kuhn@lmu.de

P. Vogt
Institute of Teacher Training (ILF) Mainz, Mainz, Germany
e-mail: vogt@ilf.bildung-rp.de

Fig. 67.1 Image of the diffraction pattern on the cell phone display

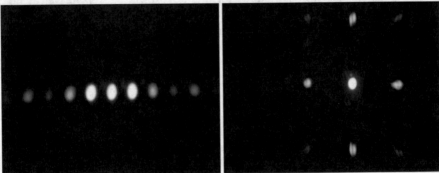

Fig. 67.2 Diffraction patterns produced with an infrared remote control and transmission gratings, recorded with a cell phone camera

67.2 Quantitative Experiments

In order to determine the wavelength of the infrared radiation produced by a remote control device, the experimental setup is adjusted as in Fig. 67.3a. A ruler, oriented perpendicular to the grating lines and placed just above or below the radiation source, serves as a measuring scale.

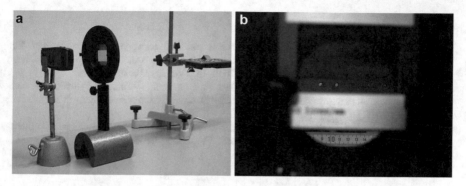

Fig. 67.3 Determining the wavelength of infrared radiation of a remote control: (**a**) experiment setup; (**b**) image of diffraction pattern and ruler

Fig. 67.4 Geometry of the experimental setup

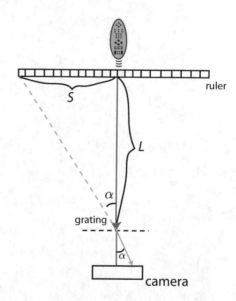

The grating and cell phone must be positioned so that the infrared radiation hits the grating and the diffraction pattern can be seen, together with the ruler, on the cell phone screen (Fig. 67.3b). After the photograph is taken, the wavelength of the infrared radiation can be determined using the diffraction grating equation, $d \sin \alpha = n\lambda$, where d is the grating spacing, n is the order number, and λ is the wavelength. Since the angle α is small, we can write $\sin \alpha \approx \tan \alpha = S/L$ (Fig. 67.4).

A typical result, using a BenQ Model CT050606055 remote control, a grating having 80 lines/mm ($d = 1.25 \cdot 10^{-5}$ m), and $L = 0.77$ m, is $S = 11.5$ cm (for $n = 2$), which results in a wavelength $\lambda = 930$ nm. This value lies within the usual wavelength range for the infrared diodes used in remote controls (900 nm – 1200 nm).

References

1. Kuhn, J., Vogt, P., Müller, S.: Neue Experimente mit dem Handy im Physikunterricht. In: Höttecke, D. (ed.) Naturwissenschaftliche Bildung als Beitrag zur Gestaltung partizipativer Demokratie: GDCP-Jahrestagung in Potsdam 2010. LIT-Verlag, Münster (2011)
2. Kuhn, J., Vogt, P., Müller, S.: Handys und Smartphones – Einsatzmöglichkeiten und Beispielexperimente im Physikunterricht. Praxis der Naturwissenschaften – Physik in der Schule. **7**(60), 5–11 (2011)
3. Catelli, F., Giovannini, O., Bolzan, V.D.A.: Estimating the infrared radiation wavelength emitted by a remote control device using a digital camera. Phys. Educ. **46**, 219–222 (2011)

Characterization of Linear Light Sources with the Smartphone's Ambient Light Sensor

Isabel Salinas, Marcos H. Giménez, Juan A. Monsoriu, and Juan C. Castro-Palacio

The smartphone's ambient light sensor has been used in the literature to study different physical phenomena [1–5]. For instance, Malus's law, which involves the polarized light, has been verified by using simultaneously the orientation and light sensors of a smartphone [1]. The illuminance of point light sources has been characterized also using the light sensor of smartphones and tablets, demonstrating in this way the well-known inverse-square law of distance [2, 3]. Moreover, these kinds of illuminance measurements with the ambient light sensor have allowed the determination of the luminous efficiency of different quasi-point optical sources (incandescent and halogen lamps) as a function of the electric power supplied [4]. Regarding mechanical systems, the inverse-square law of distance has also been used to investigate the speed and acceleration of a moving light source on an inclined plane [5] or to study coupled and damped oscillations [6]. In the present work, we go further in presenting a simple laboratory experiment using the smartphone's ambient light sensor in order to characterize a non-point light source, a linear fluorescent tube in our case.

I. Salinas (✉) · M. H. Giménez · J. A. Monsoriu
Centro de Tecnologías Físicas: Acústica, Materiales y Astrofísica, Universitat Politècnica de València, Camì de Vera s/n, València, Spain
e-mail: isalinas@fis.upv.es; mhgimene@fis.upv.es; jmonsori@fis.upv.es

J. C. Castro-Palacio
Centro de Tecnologías Físicas: Acústica, Materiales y Astrofísica, Universitat Politècnica de València, Camì de Vera s/n, València, Spain

Imperial College, London, UK
e-mail: juancas@upvnet.upv.es

68.1 Basic Theory

The smartphone's ambient light sensor is able to measure the illuminance (E) provided by an optical source, which is defined as the luminous flux (φ) per unit area (A):

$$E = \frac{\varphi}{A}. \qquad (68.1)$$

We can consider the optical source as a point source when its size is negligible compared to the distance between the detector and the source. In this case, the emitted wavefronts can be considered as spherical surfaces of radius r centered on the point source. Thus, in mathematical terms, the illuminance can be expressed as

$$E = \frac{\varphi}{4\pi r^2}. \qquad (68.2)$$

Therefore, for point optical sources, the illuminance is governed by the inverse-square law of distance. However, if the resulting illuminance from a non-negligible size light source were to be calculated at a given distance, an integral over the actual geometry of the source would have to be performed. One way to avoid using concepts that are more complex in high school and first-year university levels is to consider a linear source as an example of a non-point source. In this case, the luminous flux is distributed over cylindrical wavefronts (Fig. 68.1), so the illuminance is characterized by the following equation,

Fig. 68.1 Cylindrical wavefronts produced by a linear light source

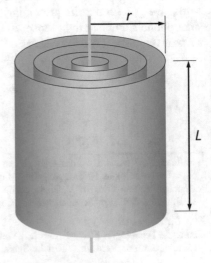

$$E = \frac{\varphi}{4\pi rL},$$ (68.3)

where r is the distance from the detector to the center of the linear source of length L. Thus, the luminance is only proportional to the inverse of the source-detector distance and not to the inverse of the squared distance. It can be noticed that Eqs. (68.2) and (68.3) are completely analogous to the equation of the electric field generated by a point charge ($E \sim 1/r^2$) or by an infinite line of charge ($E \sim 1/r$), respectively. The objective of this work is to verify experimentally the illuminance dependence, $1/r$, for linear sources using the ambient light sensor of a smartphone.

68.2   Experiments and Results

Most smartphones nowadays bear a light sensor, which allows illuminance of any light source placed nearby to be measured. Here, we will also use the light sensor but this time to measure the resulting illuminance of a light source of non-negligible size. To keep it simple, we have chosen the case of a linear source that is represented in our experiments as a conventional fluorescent tube of length $L = 120$ cm. The fluorescent tube (OSRAM T8, 36 W, 3350 lm) and the smartphone (Samsung Galaxy S7), while measuring the illuminance with the light sensor, are included in the photo of Fig. 68.2.

In order to collect the sensor data, the Physics Toolbox Suite free application for Android has been used [7]. Using this simple experimental setup (Fig. 68.2),

Fig. 68.2 Photo of the experimental setup

Fig. 68.3 Data points
collected with the light sensor.
The inverse of distance
depence of the illuminance
can be observed

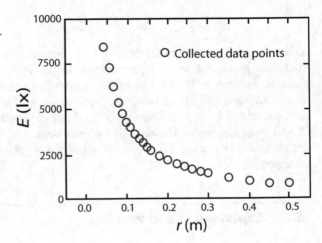

Fig. 68.4 Measured data
points (open red circles) of the
illuminance and linear fit
(black solid line) vs. the
inverse distance

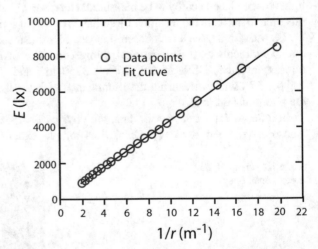

measurements of the illuminance were carried out at each distance r during 30 s. The background light was controlled such that it was kept close to zero. To perform a new measurement, the sensor was covered with an opaque black cloth until it was placed at the new position, and so on for the other measurements. The illuminance was averaged at each point over 30 s. The results showing the depence of the illuminance $E \sim 1/r$ are included in Fig. 68.3.

The luminous flux in Eq. (68.3) has been obtained by means of a linear fit using the data for E, directly measured with the light sensor vs. the inverse distance $1/r$ as

$$E = a\frac{1}{r}, \tag{68.4}$$

where a is a constant to be determined. The output of the fitting is shown in Fig. 68.4. A linear correlation coefficient of 0.9991 was obtained that shows clearly the linear dependence between the plotted variables. The resulting value of the slope,

$a = 431.3$ lx·m, was used to calculate the luminous flux $\phi = a2\pi L = 3252$ lm. This value was compared with the one reported by the manufacturer, 3350 lm. A percentage deviation of 3% was obtained, which indicates the effectiveness for teaching of the methodology presented here.This simple setup and experiment shows that the smartphone's ambient light sensor is fair enough to verify the inverse-distance law for linear sources ($E \sim 1/r$). This kind of smartphone physics experiment is being implemented with success in the first engineering courses at the School of Design Engineering, Universitat Politècnica de València, Spain.

Acknowledgments The authors would like to thank the Institute of Educational Sciences of the Universitat Politècnica de València (Spain) for the support of the Teaching Innovation Groups MoMa and e-MACAFI.

References

1. Monteiro, M., Stari, C., Cabeza, C., Martí, A.C.: The polarization of light and Malus's law using smartphones. Phys. Teach. **55**, 264 (2017)
2. Viera, L.P., Lara, V.O.M., Amaral, D.F.: Demonstration of the inverse square law with the aid of a tablet/smartphone. Rev. Bras. de Ens. de Fis. **36**, 1–3 (2014)
3. Klein, P., Hirth, M., Gröber, S., Kuhn, J., Müller, A.: Classical experiments revisited: Smartphones and tablet PCs as experimental tools in acoustics and optics. Phys. Educ. **49**, 412–417 (2017)
4. Sans, J.A., Gea-Pina, J., Gimenez, M.H., Esteve, A.R., Solbes, J., Monsoriu, J.A.: Determining the efficiency of optical sources using a smartphone's ambient light sensor. Eur. J. Phys. **38**, 025301 (2017)
5. Kapucu, S.: Finding the acceleration and speed of a light-emitting object on an inclined plane with a smartphone light sensor. Phys. Educ. **52**, 055003 (2017)
6. Sans, J.A., Manjón, F.J., Pereira, A.L.J., Gomez-Tejedor, J.A., Monsoriu, J.A.: Oscillations studied with the smartphone ambient light sensor. Eur. J. Phys. **34**, 1349–1354 (2013)
7. https://ogy.de/physicstoolboxsensorsuite

Part XIII

Astronomy and Modern Physics

Smartphone Astronomy

69

Marcus Kubsch and Hendrik Härtig

The quality of smartphone cameras has improved so far that it is possible to capture stars, planets, and the International Space Station (ISS). Inspired by the usage of a tablet in introductory astronomy presented by Gill and Burin [1], an approach to astronomy where students use smartphones to perform astronomical measurements themselves will be presented.

69.1 The Night Sky

If one considers that astronomical observations mark the beginning of the scientific consideration of our environment for humankind [2] and that events such as the landing of NASA's Spirit rover still produce wide echoes in the media, it appears strange that astronomy plays such a small role in physics curricula today. For example, astronomical knowledge can help to solve intriguing riddles in history, art, and literature or deepen our understanding of the latter (for examples, see the work of Olson [3] and Huth [4]). Therefore, astronomy can serve as a topic for projects across subjects and can help to get students engaged in science.

M. Kubsch (✉)
Leibniz-Institute for Science and Mathematics Education, Kiel, Germany
e-mail: kubsch@leibniz-ipn.de

H. Härtig
Universität Duisburg-Essen, Campus Essen, Fakultät für Physik, Didaktik der Physik, Essen, Germany
e-mail: hendrik.haertig@uni-due.de

69.2   Observing the ISS

The technical limitations of smartphones consequently limit the choice of objects
that can be observed and the types of measurements that can be performed with an
appropriate accuracy. The International Space Station is an ideal object to start with
due to its brightness (many smartphones can identify it) and speed (two observations
possible within 90 min; the transit lasts no longer than 6 min, which is consistent
with the maximal film duration of most smartphones). As depicted in Fig. 69.1, the
ISS orbits Earth in a nearly circular orbit. Therefore, one can calculate its orbital
velocity v_{ISS} via the following relation

$$\frac{2 \cdot \pi \cdot h_{\text{ISS}}}{360°} \cdot \omega_{\text{app}} = v_{\text{ISS}}, \tag{69.1}$$

where h_{ISS} is the orbital height of the ISS [5] and ω_{app} the measured angular velocity
of the ISS in degrees per second. The orbital height of the ISS is measured from
Earth's center, and so students should be led to understand that it can be estimated as
just slightly more than Earth's radius (otherwise it would not be readily visible from
Earth). However, it does change slightly over time (e.g., to save fuel, evade debris,
etc.), so it is also useful to obtain a more accurate value from Ref. [5]. We can
measure ω_{app} by filming a transit of the ISS through the field of view (FOV) of a
smartphone camera. For many smartphones the FOV of the camera can be found on
the Internet. However, measuring the FOV is a fairly easy task that can also be
performed by the students themselves. A possible setup for this is depicted in
Fig. 69.2, where the camera is aligned to two reference points A and B. After
measuring the distances y and x, the FOV can be calculated via the following
relation:

Fig. 69.1 Overview: note
that the relative distance from
Earth's surface to the ISS is
exaggerated

Fig. 69.2 Measuring
the FOV

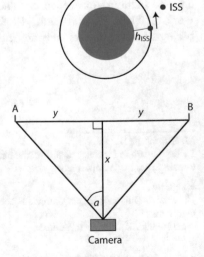

$$\text{FOV} = 2\alpha = 2 \cdot \arctan\left(\frac{y}{x}\right). \tag{69.2}$$

It is important that A, B, and the camera remain on one level, and that x and y are perpendicular to each other. For this and the later filming of the transit, it is highly advisable to use a tripod.

After a transit is filmed, ω_{app} can be calculated via the following formula:

$$\omega_{app} = \frac{\text{FOV}}{t}, \tag{69.3}$$

where t is the time that passed between the ISS's entering and leaving the FOV. NASA offers a web service [6] to find out at which time the ISS is visible at a certain location. When this method was tested with an iPhone 5 (Fig. 69.3), during two consecutive transits, the velocity could be calculated within 90% accuracy of the mean orbital velocity of the station of 7666 ms^{-1}, as presented in Table 69.1. This seems to be a fine result given errors due to weather conditions, problems with the tripod mount, and ± 1 s and $\pm 0.5°$ observational error. One of the benefits of this method is that there is no need for additional applications, i.e., the data can be evaluated directly after the measurement.

69.3   Conclusion

The methods presented in this chapter are not only suitable for secondary education but could also be used in a hands-on introductory astronomy class in college and university. It should have become clear throughout this work that smartphones and the oft-neglected topic of astronomy offer great chances to get students interested in physics and enrich classes in other subjects.

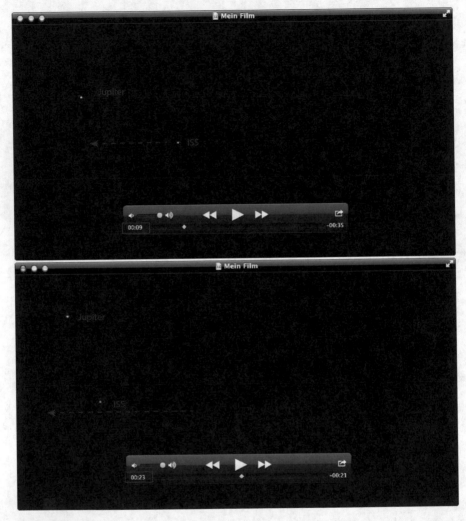

Fig. 69.3 The ISS crosses the FOV from right to left below Jupiter (both appear only as white dots). The two pictures are unaltered stills from a video (shot with an iPhone 5 in an urban area) taken 14 s apart as can be read off from the timecode

Table 69.1 Data from two consective transits filmed with an iPhone 5

Transition time (s)	FOV (°)	v_{ISS} (ms^{-1})
6 ± 1	6.25 ± 0.5	7700 ± 1000
9 ± 1	8.75 ± 0.5	7200 ± 1200

References

1. Gill, R.M., Burin, M.J.: Enhancing the introductory astronomical experience with the use of a tablet and telescope. Phys. Teach. **51**, 87–89 (2013)
2. Backhaus, U.: Astronomie im Physikunterricht. In: Physikdidaktik, pp. 485–506. Springer (2007)
3. Olson, D.W.: Celestial Sleuth – Using Astronomy to Solve Mysteries in Art, History and Literature. Springer (2014)
4. Huth, J.E.: The Lost Art of Finding Our Way. Harvard University Press (2013)
5. http://www.heavens-above.com/IssHeight.aspx
6. http://spotthestation.nasa.gov

Determination of the Orbital Inclination of the ISS with a Smartphone

70

Julien Vandermarlière

Space conquest is a very attractive subject. Who didn't dream about being an astronaut even once in his or her life? Recently, French astronaut Thomas Pesquet spent almost 200 days in the ISS. As a French science teacher, I saw a chance to get my students interested in a lot of topics. In this chapter we will expose a method that can be used to determine the orbital inclination of the ISS using a smartphone. It can be performed by even the youngest students. Moreover, the results of this experiment will allow you to answer a question that one of them asked me: "Why does the ISS track looks like a sine wave on a world map?"

70.1 Background

By definition, the orbital inclination of a satellite is the angle between the equatorial plane of Earth and its orbital plane (Fig. 70.1).

The orbital inclination of the ISS is 51.6°. This is a compromise among a few factors. Indeed, launching a rocket is extremely expensive and requires a lot of energetic consumption. In order to minimize this consumption, you'd better launch the satellite to the east in order to take advantage of Earth's rotation. Therefore, 46° would have been the most economic value since it is the value of the launch site's latitude in Baikonur, Kazakhstan. But the United States and Russia decided to choose a slightly higher value to avoid the overfly of China during launch. This particularly makes sense in case of an aborted or failed launch! Moreover, this high inclination allows the space shuttle to overfly 75% of the planet and 95% of its inhabited lands. The value of 51.6° is therefore a compromise between economic, political, and scientific reasons. Further interesting information can be found at the NASA website [1].

J. Vandermarlière (✉)
Lycée Jean Lurçat, Perpignan, France
e-mail: julien.vandermarliere@ac-montpellier.fr

© The Author(s), under exclusive license to Springer Nature Switzerland AG 2022
J. Kuhn, P. Vogt (eds.), *Smartphones as Mobile Minilabs in Physics*,
https://doi.org/10.1007/978-3-030-94044-7_70

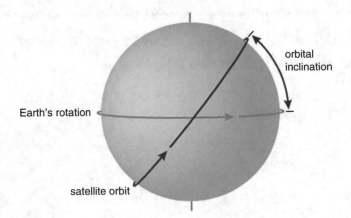

Fig. 70.1 Orbital inclination (Credit NASA, illustration by Robert Simmon)

In this experiment, a simple way to determine its value using basic triangulation is shown. Triangulation was often used in the eighteenth century when scientists were measuring the world. For instance, Delambre and Méchain, two French scientists, used it while they were measuring the distance between the cities of Dunkerque and Barcelona. This seven-year journey led to one of the first definitions of the meter [2]. Nowadays, this is the method used by GPS satellites to locate yourself!

70.2 The Experiment

It has already been reported in this column that a smartphone's sensors are good enough to track the ISS [3]. In this experiment, a Samsung S5 and three android apps are used. Dioptra™ [4] is a camera position and angle measurement tool. It allowed us to record an ISS transit and measured its azimuth and altitude vs. time. Heavens-Above [5] is a very helpful tool to plan an observation of a transit, and Mobizen [6] is a screen recorder. Finally, Stellarium7 Astronomy Software [8] was used to compare the experimental values and the theoretical ones.

Unfortunately, conditions were pretty bad: I was in town with a partially cloudy sky, and it was only the beginning of dusk (Fig. 70.2). Nevertheless, it was good enough to extract interesting data. In the video, two moments were chosen, one at the beginning and one close to the end. They were chosen because the ISS was on the center of the target. According to Heavens-Above, during this transit the maximum elevation of the space shuttle was 38°, and its magnitude was −3.1.

To construct a rough trajectory of the ISS, we used triangulation. Only the azimuths of the ISS from two different locations were needed (Table 70.1). The experiment was conducted Feb. 7, 2019, in Cabestany (42.7°N, 2.9°E). The second place chosen was Montpellier (43.6°N, 3.9°E). These are two cities located in the south of France, distant by approximatively 125 km as the crow flies. The data for

UTC: 2019.02.07T17:41:15Z
Lat, Lon: 42,678343, 2,936773
Alt: 73m MSL WGS84
CEP. 3m

Azimuth and Bearing
340° N69W

+5°
+15°
1,1°
29,1° +30 00:25
0°
+45°

NW

Recorded by Mobizen

Fig. 70.2 Screenshot of the recorded transit

Table 70.1 Experimental and theoretical azimuths of the ISS

Local time	Experimental azimuths (Cabestany)	Theoretical azimuths (Cabestany)	Theoretical azimuths (Montpellier)
18 h 41 min 15 s	340°W	343°W	334°W
18 h 41 min 51 s	3°E	5°E	356°W

Montpellier were simulated with Stellarium.4 Experimental and theoretical values differ by only 2° or 3°, which is surprisingly good.

Now, you just have to draw a few lines using Google Earth and you are almost done (Fig. 70.3)!

As this experiment is intended to be done by young students, let's keep it as simple as possible: draw two lines per location according to the azimuths in Table 70.1. Their intersections create two points that support the ISS trajectory.

Extend the trajectory line till the equator, then measure the inclination (Fig. 70.4). The value shown by Google Earth is the azimuth, calculated from the north, but the orbital inclination reference is the equator plane. It is therefore necessary to perform a very simple calculation to reach our goal:

$$90° - 37° = 53° \tag{70.1}$$

The theoretical orbital inclination of the ISS is 51.6°. Only 1.4° separate the experimental and theoretical values, which seems to be a pretty good result!

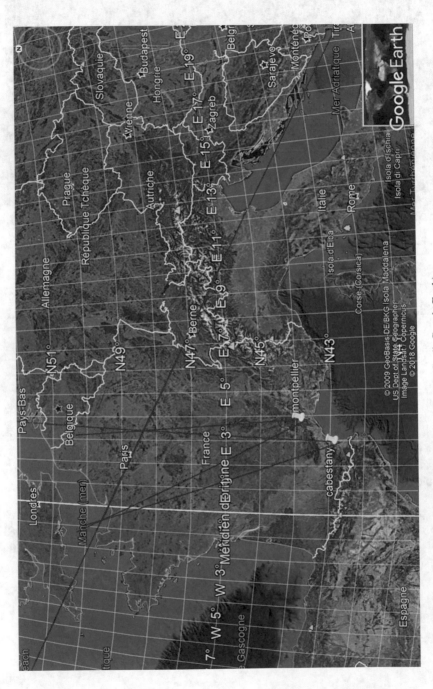

Fig. 70.3 Triangulation methodology to build the trajectory of the ISS (Credit Google Earth)

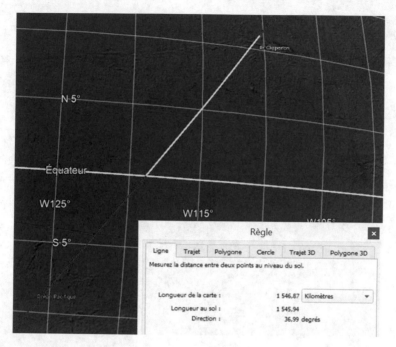

Fig. 70.4 Azimuth of the ISS trajectory measured at the equator: 37°E (Credit Google Earth)

70.3 Comments

Using smartphones and triangulation to determine the ISS trajectory is very simple and effective. We can suppose that all the other characteristics (speed, altitude) can be calculated this way, even if it could be necessary to perform more complex calculations because of the rotundity of Earth. It is also interesting to note that this work can answer the question my student asked me. At the end of the video [7], we can see the complete trajectory projected on Earth. By watching it, it's now easy to understand why it looks like a sine wave on a world map. It only remains to understand that Earth rotates under this red line. The trajectory of the ISS, very complicated at first, is now very easy to understand!

References

1. Catalog of Earth Satellite Orbits, NASA. https://earthobservatory.nasa.gov/features/OrbitsCatalog
2. Alder, K.: The Measure of All Things: The Seven-Year Odyssey and Hidden Error that Transformed the World. Free Press (2003)
3. Meißner, M., Härtig, H.: Smartphone astronomy. Phys. Teach. **52**, 440–441 (2014)
4. https://ogy.de/dioptralite

5. https://ogy.de/heavensabove
6. https://ogy.de/mobizenbildschirmaufzeichnung
7. iss transit 02 2019. https://youtu.be/wn7AvntJYWA
8. https://stellarium.org/

Adaptation of Acoustic Model Experiments of STM via Smartphones and Tablets

71

Michael Thees, Katrin Hochberg, Jochen Kuhn, and Martin Aeschlimann

The importance of Scanning Tunneling Microscopy (STM) in today's research and industry [1] leads to the question of how to include such a key technology in physics education. Manfred Euler has developed an acoustic model experiment to illustrate the fundamental measuring principles based on an analogy between quantum mechanics and acoustics [2, 3]. Based on earlier work (Chaps. 44 and 47) [4, 5] we applied mobile devices such as smartphones and tablets instead of using a computer to record and display the experimental data and thus converted Euler's experimental setup into a low-cost experiment that is easy to build and handle by students themselves.

71.1 Experimental Background

The fundamental physical process of STM is the quantum mechanical tunnel effect [6]: Applying a bias voltage between two conductive objects, i.e., an atomically sharpened tip and the sample surface [7], results in a small electric current even before they are in contact. Measuring this tunneling current as a function of the bias

M. Thees (✉) · K. Hochberg
Department of Physics/Physics Education Group, Technische Universität Kaiserslautern, Kaiserslautern, Germany
e-mail: theesm@physik.uni-kl.de

J. Kuhn
Ludwig-Maximilians-Universität München (LMU Munich), Faculty of Physics, Chair of Physics Education, Munich, Germany

M. Aeschlimann
Department of Physics, Ultrafast Phenomena at Surfaces, Technische Universität Kaiserslautern, Kaiserslautern, Germany

© The Author(s), under exclusive license to Springer Nature Switzerland AG 2022
J. Kuhn, P. Vogt (eds.), *Smartphones as Mobile Minilabs in Physics*,
https://doi.org/10.1007/978-3-030-94044-7_71

voltage and the tip position during different operating modes provides information about the surface appearance (topography mode) and chemical structures like the local electronic density of state (LDOS) of the sample (spectroscopy mode) [6].

Regarding topography mode, there are two basic ways to scan the sample [6]: During the "Constant Current Mode" (CCM), a constant bias voltage is being applied while moving the tip step-by-step over the sample with respect to a constant measuring signal. To match this set point, the tip-sample distance is adjusted, revealing the topography. During the "Constant Height Mode" (CHM), the tip is moved over the sample without adjusting the height. Consequently, the instantaneous measuring signal provides the information about the topography.

However, there are some difficulties: If the surface consists of just one element, the topography relates to the "real" position of the atoms. Other elements might not be detected due to their different LDOS causing another behavior of the tunneling current as a function of the bias voltage and the tip-sample distance. This behavior is investigated during spectroscopy mode. By measuring the dependencies mentioned above, it is possible to specify the involved elements at a fixed viewpoint [6].

71.2 Analogies

Based on the work of Euler [2, 3], we used the analogies in Table 71.1 to adapt the model experiment.

The plastic bottles act as acoustic resonators, representing the (electronic) surface configuration of different atoms or molecules. Hence, the measurable data relates to the existence of acoustic resonators [2] and only one particular condition to the "real" position of an atom.

71.3 Adapted Experimental Setup

To set up the experiment, just a few everyday materials are needed: two smartphones or tablets (we used Apple iPod, 5th generation), two headsets, and some plastic bottles (e.g., yogurt drinks; Fig. 71.1). In addition, we used the following applications: Signal Generator, Noise [8], and SpectrumView [9].

Table 71.1 Analogies between quantum mechanics and acoustics

Quantum Mechanics	Acoustics
Matter wave	Sound wave
Tunnel effect	Acoustic resonance
Bias voltage (= energy of electrons)	Frequency
Tunneling current	Sound pressure level (microphone)
Measuring amplification	Sound pressure level (headphones)

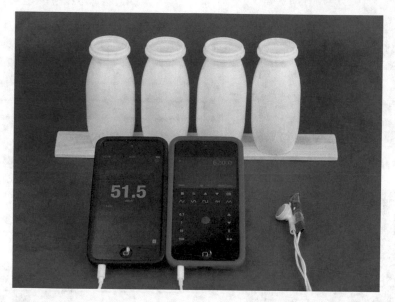

Fig. 71.1 Photo of materials and prepared tip

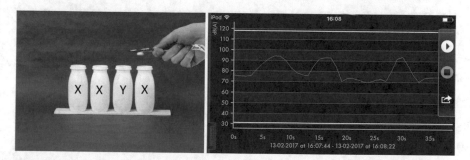

Fig. 71.2 Experimental setup of pre-experiment (left) and screenshot of measurement provided by Noise app (right). The bottle types are marked

Each mobile device is connected to one headset. To build the measuring head (tip), the headphone of device A is taped together with the microphone of the headset of device B.

We added water to some of the bottles (up to a quarter of the bottle) to manipulate their resonance frequency in order to receive two different types of acoustic resonators (X and Y). The measurement signal is the sound pressure level (SPL) detected by the Noise app on device B. On device A, we applied a sine wave with the app Signal Generator using the resonance frequency of bottle type X. With this setting, it is possible to detect type X but not type Y.

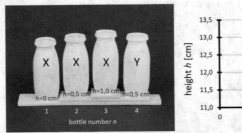
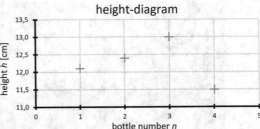

Fig. 71.3 Experimental setup (left) and height diagram (right) of CCM. Constant SPL: 105 dB

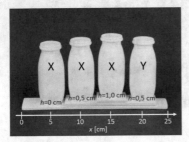
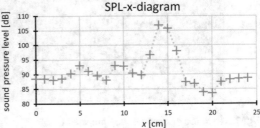

Fig. 71.4 Experimental setup (left) and SPL-x diagram (right) of CHM. Ground-tip distance: 13 cm. Step width: 1 cm

Qualitative Pre-Experiment

The bottles are mounted on the same height level. The tip is slowly moved over the surface by hand. To display the data, the view of the SPL over time provided by Noise is to be chosen. The results showed that type X was detected, while type Y was not (Fig. 71.2).

During topography mode, the bottles are mounted at different height levels, forming the sample surface structure. To receive quantitative data, the surface is scanned systematically. The height of the ground-tip distance and the SPL are measured.

Quantitative Illustration of the Operating Modes

Constant Current Mode: During the scan, the tip is placed just above the middle of the bottle opening and the ground-tip distance is adjusted with respect to a constant SPL using a ruler or tape measure (Fig. 71.3, left). This ground-tip distance over each bottle represents the topography, facing the problem that the position of type Y does not relate to its "real" position (Fig. 71.3, right).

Constant Height Mode: During a one-dimensional scan, a step width of $d = 1$ cm was chosen. The results showed that for bottle type X the measuring signal represented the topography, while bottle type Y was detected as type X with a lower height level (Fig. 71.4). To receive more accurate values, the time-related averaging offered by the Noise app can be used. Similarly, two-dimensional scans are feasible.

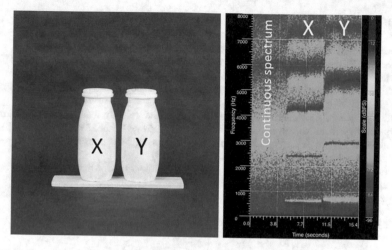

Fig. 71.5 Experimental setup of spectroscopy mode (left) and screenshot of measurement provided by SpectrumView (right). The bottle types are marked

Instead of using a sine wave and varying the acoustic frequency during spectroscopy mode, we applied a continuous acoustic spectrum (e.g., "white noise"), modulated by the frequency response of the microphone (Fig. 71.5). Displaying the received signal in a spectrogram (SPL-frequency diagram over time) provided by SpectrumView, it is possible to determine the different resonance frequencies of each bottle and to distinguish two different resonant structures, representing the different energy levels of electron bands in our analogy.

71.4   Conclusion

We have presented a low-cost model experiment based on the ideas of Manfred Euler to explore the fundamental measuring principles and operating modes of STM. With this setup, students can reproduce the measuring process using just a few everyday materials. Especially the application of smart mobile devices allows a quick and clear presentation of the received data. Regarding students' naive beliefs [2] of atoms as hard balls, the visualization and the correct interpretation of the experimental data remain main educational challenges. Considering the analogies between quantum physics and acoustics, students are able to gain ideas about the nonclassical nature of electron tunneling and the appearance of atoms and surface structures on an atomic scale.

Acknowledgment   Financial support by SFB/TRR 173 "Spin+X: Spin in its collective environment" (outreach project) of the Deutsche Forschungsgemeinschaft (DFG) is gratefully acknowledged.

M. Thees et al.

References

1. Morita, S. (ed.): Roadmap of Scanning Probe Microscopy. Springer, Berlin (2007)
2. Euler, M.: Near-field imaging with sound. An acoustic STM model. Phys. Teach. **50**, 414–416 (2012)
3. Euler, M.: The sounds of nanoscience. Acoustic STM analogues. Phys. Educ. **48**(5), 563–569 (2013)
4. Hirth, M., Kuhn, J., Müller, A.: Measurement of sound velocity made easy using harmonic resonant frequencies with everyday mobile technology. Phys. Teach. **53**, 120–121 (2015)
5. Monteiro, M., Marti, A.C., Vogt, P., Kasper, L., Quarthal, D.: Measuring the acoustic response of Helmholtz resonators. Phys. Teach. **53**, 247–248 (2015)
6. Voigtländer, B.: Scanning Probe Microscopy. Atomic Force Microscopy and Scanning Tunneling Microscopy. Springer, Heidelberg (2015)
7. Foster, A., Hofer, W.A.: Scanning Probe Microscopy: Atomic Scale Engineering by Forces and Currents. Springer, New York (2006)
8. This application is currently not available due to update problems
9. https://ogy.de/spectrumview